GROUND | WATER

The Art, Design, and Science of a Dry River

CONFLUENCENTER BEYOND BOUNDARIES SERIES, VOL No. 1

Javier Duran, Series Editor

GROUND|WATER

The Art, Design, and Science of a Dry River

Edited by

ELLEN MCMAHON

ANDER MONSON

BETH WEINSTEIN

CONFLUENCENTER FOR CREATIVE INQUIRY

The University of Arizona • Tucson, Arizona

Published by:
Confluencenter for Creative Inquiry

Distributed by:
The University of Arizona Press
1510 East University Blvd.
Tucson, Arizona 85721
www.uapress.arizona.edu

ISBN 978-0-8165-3023-6
Library of Congress Cataloging-in-Publication Data
available from the publisher upon request.

Project Coordinator: ELLEN MCMAHON

Editors: ELLEN MCMAHON, ANDER MONSON, BETH WEINSTEIN

Book Design: JONATHAN BLACK, KELLY LESLIE, ELLEN MCMAHON

Endpaper and Essay Photography: CAMDEN HARDY

Paper: NEW LEAF EVEREST AND MOHAWK LOOP

Printing: ARIZONA LITHOGRAPHERS, TUCSON, ARIZONA

Cover Printing: LETTERPRESS FINESSE, TUCSON, ARIZONA

Binding: ROSWELL BOOKBINDING, PHOENIX, ARIZONA

Typography: GOTHAM NARROW AND GOTHAM ROUNDED, DESIGNED BY H&FJ

CONTENTS

PUBLISHER'S NOTE

I am delighted to introduce the concrete results of one of the first Confluencenter faculty grants. This volume is not just about water; that is to say, it is not a scientific monograph analyzing a covalent bond of two atoms of hydrogen and one of oxygen. There is certainly much here for the hydrologist and geographer—but the historian and sociologist, architect and urban planner, and poet and artist are just as present. These approaches represent only some of the many that contributed their unique insight to the project. *Ground/Water* has a scope and magnitude that exceeds the majority of peer projects born out of academic institutions. Unlike many such projects, professors and researchers collaborated with students to produce knowledge and then stepped outside the bounds of their offices, classrooms, and laboratories to distribute it to the public. This was not a passive public, but a community that participated as much in the process of knowledge creation as the experts themselves.

As the chemical mechanism that allows water to form, the covalent bond is the obvious conceit to describe how this project bridges disciplines. The molecular constituents of *Ground/Water* enjoy their own independent existence but, in coming into contact, they create something greater than the sum of their parts. Nonetheless, this comparison is lacking because the project is more fluid than water. It is not covalent, but rather polyvalent; it moves in and out of manifold areas which, already exceeding the bounds of traditional disciplines, cohere in a way which illuminates all of its parts but which requires a public to give it life. Water enjoys an independent existence, but *Ground/Water* evaporates in the absence of the human element.

Thus, it is no coincidence that Confluencenter for Creative Inquiry is participating in this project. We are an interdisciplinary center supporting research and creative endeavors in the humanities, social and behavioral sciences, and fine arts. As such, *Ground/Water* not only encompasses our mission, but reaches the heights of what we believe interdisciplinary activities are capable of: innovative perspectives, creative insights, and an unprecedented level of interaction between the academic and public spheres.

The challenge of the 21st century is assumed to be assembling the vast amount of knowledge we have gathered about our planet in the past decades. However, *Ground/Water* speaks to the real challenge at hand: how can we escape the Sisyphean task of trying to parse a massive collection of data and begin to create meaning from it? This project evinces the solution, which is as simple as consciously operating outside of traditional boundaries. It does not exclude any discipline (or anyone at all) that has something to say about it. As humans we have a responsibility to be not just instruments of data collection, but to continually interpret from all points of view. It is our sincere hope that *Ground/Water* will engender similar projects, which will further erase the rigid disciplinary methodologies of yesteryear and advance a new paradigm: a poetics of experience and collaboration.

JAVIER D. DURAN

Ph.D., Director of Confluencenter for Creative Inquiry

FOREWORD

Ground/Water is an ode to a dry river, the kind of river most familiar to those who dwell in Southern Arizona. It is also an experiment in making something beautiful from something that has been desecrated. And it is a strong message about community and responsibility.

Although changes in the climate have undoubtedly contributed to the loss of flowing water in the rivers of Tucson, the primary responsibility lies with those of us who have lived here over the last 100 years. As with so many environmental issues, there is no mystery about what has caused the depletion of groundwater and the loss of flowing streams in Arizona. The physics of water behavior are well understood and modeled, even if there are some lingering questions about what occurs unseen in the aquifers under our feet. The basic facts are clear: using more water on an annual basis than is replenished is not sustainable over the long term.

The real mysteries are about human behavior. Why do we act as if our resources were infinite when they are so obviously finite? The issue is not inadequate science; it is an issue of perspective and world view. Science and technology have a major role in addressing these issues, but the larger challenge comes from a world view that does not adequately focus on sustainability of the human and environmental systems on which we depend.

But *Ground/Water* is also about affirmation. There is wisdom as well as beauty in this book, which starts from the premise that we do have choices and that the future brings great promise... and as the editors acknowledge in the following introduction, an important function of the arts is "to enable us to interpret what we see, clarify how we feel, determine what matters, and inspire us to act."

On page 98 of this volume, contributor Christiana Caro writes "The River, even dry, also has stories, also renders a line to be followed, and then too disappears before filling, briefly, then emptying again." The winter rains and summer monsoons bring us opportunities for renewal and inspiration. They bring us hope as well as briefly flowing rivers that recharge our spirits and our aquifers. They bring us an annual opportunity to examine our own roles in this desert ecosystem and to individually and collectively invest in the social and natural systems that support us.

KATHARINE L. JACOBS

Director of the National Climate Assessment

ACKNOWLEDGEMENTS

The editors would like to thank Confluencenter for Creative Inquiry for their generous funding of both the collaborative faculty research project "Parallel Play: Interdisciplinary Responses to a Dry River" and the production of *Ground/Water: The Art, Design, and Science of a Dry River.* Special thanks go to Javier Duran, Director of Confluencenter, and Kathryn Conrad (Director) and Kristen Buckles (Acquiring Editor) of The University of Arizona Press for their support in publishing and distributing this volume.

Financial support was also provided by the Water Sustainability Program of The University of Arizona Water Resources Research Center for the production of banners and cards for *What's So Important About Water?* by Jessica Gerlach. Program Coordinator Jacqueline Moxley and Ph.D. candidate in Soil, Water and Environmental Science Mónica Ramirez-Andreotta were particularly instrumental in creating the connections that led to this science translation funding being granted to a graduate student in the School of Art.

We are appreciative of all of the people who worked with faculty and students individually and in the context of the interdisciplinary seminar "Critical Issues in Art + Design" and architectural design studio "Water+." Our understanding of Tucson's water was enhanced by: Wally Wilson, Chief Hydrologist, Terry Miley, Water Resources Manager, and Bruce M. Prior, Hydrologist, with the City of Tucson Water Department; Stanley A. Leake and Fred D. Tillman, Research Hydrologists, U.S. Geological Survey; and Emily Brott, Southern Arizona Project Manager, Sonoran Institute. We are grateful to Landscape Architecture professor Margaret Livingston for leading our field trip to the Rillito, and architectural critics and faculty members Christopher Domin, Michael Kothke, Annie Kurtin, Bill Mackey, Wil Peterson, Gundula Proksch (University of Washington), Chris Trumble, and directors Lauri Johnson and Robert Miller.

People from a number of disciplines shared their research and creative work during the course of this project including: Geography Ph.D. students Jacob C. Miller and Sunčana Laketa, Jamie A. Lee of visionaries fieldworks inc., and Steev Hise of Pan Left Productions. Our collaborators, the Rillito River Project, provided inspiration and connection to a large community supportive of art–science collaborations. In particular, we thank Ellen Skotheim for her vision, Jennifer Powers-Murphy for her publicity skills, and John Newman for his technical know-how. Thank you also to Bat Conservation International for providing imagery of bats in flight for the visual identity of Bat Night 2010.

Our most persistent science partner, from his enthusiasm about the initial drafts of the grant proposal "Parallel Play," through his participation in the interdisciplinary seminar, to the final edits of this volume, has been Gregg Garfin, Deputy Director of Science Translation and Outreach in The University of Arizona's Institute of the Environment. We are indebted to him for sharing his knowledge of local water and climate and his appreciation of the importance of the arts in changing people's attitudes and behaviors. He is responsible for substantially increasing the scientific accuracy of the text but not for any errors or oversights, which are the sole responsibility of the contributors.

For supporting faculty in cross-disciplinary activities we'd like to acknowledge our home departments: the School of Art, the College of Architecture and Landscape Architecture, the Creative Writing Program, the Department of English, the School of Geography and Development, and the Institute of the Environment. Special thanks go to the School of Art for administering the grant, and the Drachman Institute and UA Downtown for providing the venue for the exhibition "Parallel Play" in October of 2011.

We are deeply appreciative of our students for their diligent research and commitment to their individual creative practices, which are evident in the brilliant and beautiful works that grace these pages. Jonathan Black deserves special mention as he started as a graduate assistant for "Parallel Play" in 2010 and continued working on the design of this book months after his graduation from the School of Art MFA program in 2012. Without his perfectionism, obsessive interest in typography and design, and dedication to the project, this book would not be the lovely object it is.

For collaborations to be enjoyable and worthwhile great colleagues are required, and we have them in Nathaniel Brodie, Alison Hawthorne Deming, Allison Dushane, Logan Phillips, Paul Robbins, and Gregg Garfin, whose contributed poems and essays have expanded the scope and depth of our thinking and our work. Without Kelly Leslie's book design expertise and Cybele Knowles' intelligent and keen copy editing, this book would be far less consistent, incisive, and accurate. Their support at the later stages of the project were essential to bringing this project to fruition. Thank you also to Matthew Conley for his help in clarifying the writing and to Conley and Yvonne Ervin, Linda Dobbyn, and Daniel and Cindy Brocious for swooping in at the last stages to read the final drafts.

Every effort was made to reduce the environmental impact throughout the design and production of this book. The endpapers are printed on Mohawk Loop and the inside pages are printed on New Leaf Everest, both 100% postconsumer recycled papers manufactured with renewable energy. Thank you to New Leaf for reducing by 20% the cost of the paper. Arizona Lithographers was chosen specifically for their dedication to environmental responsibility (their plant is 100% wind powered) and great appreciation is owed to them for their patience, guidance, and generous support of the project. The inside pages are printed using vegetable-based inks. The covers are printed on binder's board made of 100% recycled fiber. The entire book is printed on uncoated paper, which ensures efficient recyclability at the end of its life.

The photographs on the endpapers and on pages X, 6, 30, 46, and 58 were taken by Camden Hardy, a member of the interdisciplinary seminar in 2010 and a graduate of the School of Art MFA program in 2012. Hardy's work is inspired by the idea that bonds form between people through the shared experience of material objects spanning across time. In his artist statement Hardy writes:

> As social beings, we are compelled to share our experiences with others. To see evidence that another person once existed where we stand is comforting. It tells us that we are not alone—that others have found meaning in their surroundings.

The essence of Hardy's artistic intent has manifest in his significant contribution to *Ground/Water*, this shared object, which has formed bonds among its creators. We hope it will continue to have a similar effect on others throughout its lifetime.

ELLEN MCMAHON, ANDER MONSON, BETH WEINSTEIN

Introduction

ELLEN MCMAHON, ANDER MONSON, BETH WEINSTEIN

The starting point was the dry bed of an urban river, catalyzing the desire to understand local water and its relationship to global climate. As artists, architects, and writers, we approached these issues through our own creative practices, using data and resources from the natural and social sciences in supporting roles. Our goal was to cross-pollinate the objective procedures of science, the more subjective processes of art, and the hybrid methodologies of design. Collaborations within The University of Arizona and with the Rillito River Project, a Tucson-based arts group, resulted in the following confluence of public events, photographs, graphic design, architectural proposals, artist books, essays, and poems. Together these works build a case for employing diverse methods of inquiry, representation, and communication for the purposes of understanding our experience, imagining new possibilities, and responding to increasingly complex environmental challenges.

SITE SPECIFICS

Throughout the desert city of Tucson, Arizona, bridges cross wide, dusty ditches with names such as the Santa Cruz River, Cienega Creek, and the Pantano Wash. Water is only visible in these 'rivers' during brief periods of winter rain and in the summer monsoon season, when they act as drainage channels for massive amounts of roiling muddy water. As in much of the rest of the arid southwestern United States, population growth, sustained drought, and increased temperatures associated with global climate change have put substantial pressure on Tucson's water supplies. Though residents have been pumping their groundwater unsustainably since the 1880s, several organizations promoting water conservation and environmental restoration, assisting communities with water management and policy, and conducting and supporting scientific research, have contributed to the city's current proactive stance towards water.

The dramatic environment, clear light, and multicultural identity of Tucson have drawn artists, designers, and writers over many generations to the city and the surrounding Sonoran Desert. These communities have generated and nurtured distinctive cultural and natural histories of the region, utopian building projects, site-specific land art, and intimate works inspired by solitary experiences within the unique landscape. Additionally, The University of Arizona is an internationally recognized center for research concerning arid lands, water resources, and climate change. The significance of local water, the expertise in the arts and sciences, the culture of collaboration within the university, and the financial support for multidisciplinary research helped create the focus, motivation, and opportunity that culminated in *Ground/Water*.

RILLITO RIVER PROJECT COLLABORATIONS

The first wave of projects began in 2009 with an invitation to University of Arizona faculty from the Rillito River Project to collaborate on Bat Night 2010. This annual event, inaugurated in 2008 by the Rillito River Project, is staged in one of Tucson's dry riverbeds adjacent to the bridge of a busy street. Bat Night combines elements of art and science to educate a large public audience about the migratory bat colony roosting under the bridge and global environmental issues related to the river's vulnerable ecology.

Bat Night 2010 collaborations between faculty, students, and the Rillito River Project resulted in a series of projects beginning with the poem "Rillito" by Creative Writing professor Alison Hawthorne Deming and "Dry River City" written by Nathaniel Brodie, who was then studying with Deming as a creative writing MFA student. Professors Ellen McMahon (Art and Graphic Design) and Beth Weinstein (Architecture) worked with four students on projects related to Bat Night 2010. Fifth-year architecture students Nicole Sweeney and Jennifer Heinfeld proposed research-based ephemeral installations. Sweeney's proposal for a didactic installation titled *waterBar* emerged from research on structures and infrastructures that contain and channel water. Heinfeld's proposal, *Confluence of Boundaries*, drew from research concerning the Rillito as a geographical feature both dividing and joining the city. The spatial strategy in Heinfeld's project was reinterpreted to create the built environment, seating, stage, and children's play area that was ultimately realized for Bat Night 2010.

Graphic design MFA student Jessica Gerlach researched the effects of groundwater depletion on the biotic communities of the Rillito and created *What's So Important About Water?*, an interactive public art piece central to the event. Jonathan Black, also an MFA student in graphic design, created the visual identity for the event featured on posters, invitations, and children's sand pails. Bat Night 2010 attracted nearly 5,000 people into the dry Rillito riverbed. All of the aforementioned projects, in addition to a brief history of the Rillito River Project and the poem "Canto Rillito" written by Logan Philips for Bat Night 2009, can be found in this volume.

PARALLEL PLAY: FACULTY RESEARCH PROJECT

The next phase began in the spring of 2010 when McMahon, Weinstein, and Creative Writing professor Ander Monson received funding from the University Office of the Provost for the project "Parallel Play: Interdisciplinary Responses to a Dry River." Monson undertook a research project that resulted in a site-specific installation in a sewer culvert, a video titled *Take Me to the River*, and the essay published here, "I in River." Interested in bringing collaborative research into the classroom, McMahon and Weinstein co-convened an interdisciplinary seminar of art, design, and architecture students interested in addressing environmental issues through their work. Seminar participants expanded their knowledge of water through readings, discussions, and presentations by guest speakers from hydrology, geography, film, literature, and creative writing. In addition, students researched the substantial body of art, design, and architectural projects responding to environmental problems and produced visual essays synthesizing and representing their findings. Supported by a series of group critiques, each student developed a final project ranging from intercontinental comparisons of arid regions and river systems to personal and empirical reflections on the local arid landscape.

Four MFA students brought artistic research methods based on direct experience to bear on the topic at hand. John Gialanella's *REF, YOOS River* is an interactive kit containing a handmade artist book and a collection of discarded man-made objects found during his travels in the Rillito riverbed. Camden Hardy walked all 22 miles of the Pantano Wash creating a videorecording, notes of his experience, lists of objects found in the wash, and photographic artifacts such as pinhole negatives and photograms to signify the end of one mile and the beginning of the next. Daniel Cheek's series of contact prints made from 8" x 10" negatives, *Sweetwater Wetlands, 2010*, contrasts the lush vegetation at the city's Sweetwater Wetlands Recharge Facility with subtle signs of human control and management of water. Robert Long explored the dry bed of the Rillito as well as the Sweetwater Wetlands with a small handmade journal, recording his surroundings and his encounters with small creatures in finely detailed drawings and texts.

Within the framework of the seminar, municipal specialists, community activists, members of environmental organizations, and faculty members from the sciences and humanities gathered to share expertise and experience. One of the seminar's guest speakers, Tucson Water's Chief Hydrologist Wally Wilson, was a major contributor to the student project, *Tucson's Water Future.* Two faculty consultants to the "Parallel Play" project have contributed essays for this volume. Climatologist Gregg Garfin's essay "Desert Water: Paradoxes and Trade-Offs" provides a historical overview of and context for the region's climate, hydrology, and water policy. Geography professor Paul Robbins examines the differences between an idealized protocol of interdisciplinary research and one that might sustain communities of conversation around a common subject in his essay, "Talking through Objects: Multidisciplinary Dialogues with 'Things.'" English Literature professor Allison Dushane, one of the seminar's guest speakers, traces the roles of science and art in the construction of contemporary concepts of "nature" through seminal works of Romantic literature in her essay, "Imagination and Invention: Romanticism and the Life of a Dry River."

> The arts enable us to interpret what we see, clarify how we feel, determine what matters, and inspire us to act.

As Deputy Director for Science Translation and Outreach at The University of Arizona's Institute of the Environment, Gregg Garfin's role is to communicate key climate, environmental, and resource management issues to policy makers and stakeholders in the Southwest. He writes of his experience in the seminar:

> I am struck by not just the multiple ways of knowing our environment expressed in the projects, but also by the facility with which contributors transformed messages from the stilted language of the scientific literature that they consulted into rich and multifaceted vocabularies that can be grasped by those interested, but inexpert, in the subject matter. My participation in the seminar drove home to me the value of art in elucidating and explaining complex resource management challenges. The resulting projects caused me to reflect on the insights that can be gained from an infusion of new and radically different epistemologies, the abundance and agglomeration of ideas within each of us, and the growth that we can achieve by parsing the evidence through multiple perspectives.

The wide range of people who became involved in the seminar and in later stages of this project did so because they understand the role of art and design in determining how people perceive and interact with their surroundings. Although science provides the contemporary world with its information, how we feel about that information is profoundly influenced by artistic expression and visual communication. Creating the necessary change in public attitudes about the globe's limited natural resources will require a well-considered combination of artistic and scientific means.

TUCSON'S WATER: SUPPLY AND DEMAND

Graphic design and architecture projects related to future conditions were created in the interdisciplinary seminar and in a subsequent advanced architectural design studio conducted by Weinstein titled "Water+." Projects from the seminar include Jeff Leinenveber's infographic *Tucson's Water Future*, based on a lecture by Wally Wilson, and Danielle Alvarez's proposed installation, *The Nile and Colorado Rivers*. Leinenveber's series of infographics illustrating the sources of and strategies for providing water for Tucson from 2000 to 2050 are accompanied by comments by Wally Wilson. Working within the framework of "information informing form," Alvarez's project reveals a tug of war over water across borders—between states in the United States and internationally along the length of the Nile River.

In the architectural design studio students continued to research water at local, regional, and global levels and produced proposals reimagining several of Tucson's "waterfront" neighborhoods, taking into account predictions for population growth and decreased natural resources in 2050. Works developed in the context of this studio, resulting from students' research on a particular topic such as wastewater filtration or food-production water usage, led to the development of design proposals at building or urban/infrastructural scales. *Climatic Landscaped Habitat* is Jongwoo Kim's exploration of topographic manipulations along the Santa Cruz River's edges, intended to create earth berms and water retention and recharge basins to support future housing development. Mathew Propst's careful analysis and graphic translation of the vicious energy–water loop in Phoenix—between water pumping and use, and hydroelectric energy production and consumption—led him to develop a proposal for Tucson influenced by landscape urbanism works. Propst's *Tucson + Production* is an infrastructure along the Rillito riverfront to locally produce food and facilitate groundwater recharge. *Industrial Receptacles*, Alothman's proposal for a system of structures that combine water filtration, greenhouses, and markets, similarly considers architecture as infrastructural and, as a result, a part of a larger ecosystem.

INTERIOR LANDSCAPES

This book closes with two works that explore perception and identity. Ander Monson's essay "I in River" wanders between interior and exterior phenomena—experience and memory, river and storm sewer graffiti—utilizing the form of the book by collecting around and spanning the gutter where the inside edges of the pages meet. Christiana Caro's series of photographs and accompanying text, *Past Year's Data/Future Archive 2010–11*, looks far afield to the Sky Islands, the mountaintops that surround the Tucson Basin. Caro's images of accidental moments in local museums catch them in the act of freezing and transforming the natural into the cultural.

FUTURE CONFLUENCES

Technological advances and increased specialization in all of the sciences and communications fields have led to unprecedented knowledge about almost everything, from the workings of the human mind to intimate details of the most remote places on the planet. Drowning in data, the challenge is to filter, sort, and absorb this information. This is one of the most important functions of the arts: to enable us to interpret what we see, clarify how we feel, determine what matters, and inspire us to act. Through individual and collaborative creative endeavors, and with the support of the community that formed around them, *Ground/Water* participants experienced firsthand the benefits of multidisciplinary approaches and the critical role of art and design in changing the perceptions of makers and audiences alike. We offer this volume to widen the circle.

Rillito

ALISON HAWTHORNE DEMING

You might call it a *river,*
this track of sand and scrub
that threads through the city
inviting runners, riders, taggers,
thermal gliders, trotting coyotes,
and a solitary cougar.
A pueblo of bats
nestles into bridge girders
to raise their young in the ways
of their ancestors, their sonar
touching whatever comes near,
their tongues licking bugs
out of air. You might
call it a *river* when the monsoon
brings a flood that bashes
through the stillness
that threads through the city,
the brown churn hurling
dirt and weeds and shopping carts
in its flow until it settles down and drops
whatever it has carried right there
in the spot where the flowing stops.
You might call it a *river*
unless you came to its shores
the summer when the mountains raged
with wild fire and the forest turned to char.
The rains began the healing,
but it was slow, runoff
threading through crevice, gully,
and wash, then gathering
in Aqua Caliente and Tanque Verde Wash,
then running into the stillness that
threads through the city, carrying
silt of blackened wood and rhizome
so fine it made the racing flow shine
black as liquid coal. I was there.
I saw the river at work then,
how even as it might thirst for water,
might long to break
the paving that impounds it,
how even as the river might be
a stranger to itself, lost to the city's
velocity and constant subdued roar,
how the river will rise
to the fact of weather
telling its story of beauty and loss.
You might
call it a *river*
though it looks nothing like
what you imagine a river to be,
water shining one day
black as liquid coal
and next day
becoming again
the stillness that threads through the city
where something you might
call *your life*
reveals itself to be
an eddy in the unseen flow
of all that continues.

Dry City River

NATHANIEL BRODIE

RILLITO

Meaning little river. Fair enough. The Rillito gathers out of its tributaries to run twelve miles along the southern slope of the Santa Catalina Mountains before flushing into the Santa Cruz River. Twelve miles—there it is. Except that it doesn't run. It's dry most of the year. Except that this little, twelve-mile river's watershed drains almost a thousand square miles—from the Santa Catalina to the Rincon to the Santa Rita Mountains, from Sonoran desert scrub to Rocky Mountain aspen, from three to nine thousand feet; a thousand square miles of Southern Arizona's runoff into the little, twelve-mile river. Except that this little river's main tributary, which is actually three different courses—Pantano Wash out of Las Cienegas Creek out of Gardner Canyon Creek—stretches six times the Rillito's length to its headwaters. And the story of the little river is not the twelve miles or the thousand miles but the confluences—the Rillito formed of the confluence of Pantano Wash and Tanque Verde Wash and fed—to list just those off the southern slope of the Catalina range—by Nanini Wash, Casas Adobes Wash, Citrus Wash, Coaster Wash, Pima Wash, Village Wash, Friendly Wash, Race Track Wash, Via Wash, Campbell Wash, Camino Real Wash, Finger Wash, Valley View Wash, Flecha Caida Wash. But a confluence isn't solely the merging of two flowing bodies of water, or in the Rillito's case, two nonflowing watercourses, but a concourse, a coming together of people or things, of people and things. The Rillito is a confluence—made of the actual physical junctures of tributaries and the more nebulous confluence of river and city.

CHAOS AND CAUSATION

Every instant of our individual and collective existence we affect our surroundings. We live in a world of systems and a single event, a single change, or myriad, even unrelated events can coalesce and ripple throughout an entire system, perhaps dying out, perhaps intensifying with each successive ripple. The slightest perturbation in the initial stages of a system's trajectory may lead to significantly different outcomes. And so the butterfly effect—the displaced air from a butterfly wing rippling outward, shifting weather patterns, creating a hurricane. And so chaos theory, the study of systems so sensitive to change that the slightest shift skews the entire system's dynamics. And so the Rillito River: if not a chaotic system, contains chaotic elements; if the smallest perturbation—the cow path, the death of a single cottonwood—didn't shift the system, the Rillito shifted nonetheless. Such finer distinctions are, perhaps, beside the point when considering that the earth beneath Tucson is subsiding into the void of an overtapped aquifer or that Tucson consists of a virtually contiguous slab of concrete laid over 100 square miles of alluvial plain. The story of Tucson and the Rillito is the story of cause and effect, the story of humanity's need for stability overlaid on an inherently unstable system.

ARCHEORIVERDYNAMICS

The mountain ranges ringing Tucson have been crumbling for over five million years, filling the basin between them with over five thousand vertical feet of alluvium. The accrued alluvium has been leveled into a gently sloping plain by the windshield-wiper-like action of the Rillito's ancestors. In more humid parts of the world, watercourses run perennially, and have predictable, continual adjustments between process and form—they flood at rather predictable times and the floods shape the stream rather predictably into valleys. They are stable systems, there in their little valleys. Not so the water courses along Tucson's alluvial plains. They flowed fitfully atop the basin floor, fickle to the weather—in drought they'd wither and die, in flood they'd spasm like a loose fire hose. Occasionally they'd incise, in full or in part, and within the incision they'd meander, forming cutbanks, downcutting further, until a big flood would debouch from the incision and gouge a new channel, or flow in sheets across the plain.

KNOWING A RIVER: SENSES

Tucson is a city of bridges. Fifteen bridges cross the Rillito River. We Tucsonans know the Rillito more through the brief glance at the open stretch of beige and brown and white as we bridge the river than any deeper associations or reflections between aridity and human presence. But we can scramble off the bridges; we can know this river through immersion. Every day, hundreds of Tucsonans walk, run, and bike the paths that run parallel to the Rillito. People play with their dogs and fly kites in the riverbed. This is knowledge gained through the blazing sun on the neck and forearms, the Cooper's Hawk sweeping low overhead, the rich dust in the nostrils. The satisfying crunch of gravel underfoot; the fresh anthills spilling up smoothed granite, diorite, schist, and gneiss; the clunks of concrete and gnarled rootballs woven with nests of flood debris. This is knowledge gained through sounds: Home Depot intercom, sirens, air-force bombers rumbling into Davis-Monthan Air Force Base, bat squeaks, coyote yips, the aggressive chirp of a hummingbird on a dead branch. This is knowledge through the shed detritus of the city, the flotsam and jetsam of diaper bags, pool toys, cigarette butts, bones, batteries, rebar, tires, mangled shopping carts, broken glass bottles, spectacles, unbroken plates, dead pigeons, dolls, and the thousands of plastic water containers in various states of photovoltaic decay.

ANATOMY OF A FLOOD: RAIN

It rains. Not much and not very often, about twelve inches a year in the Tucson Basin. There are winter storms, steady in rain, but most rain arrives with the monsoons, the violent summer pulsing of rain, lightning, and thunder. The monsoons roll off the Gulf so laden with rain they barely float. Rain collapses out of their black, flat bellies as though thermals were thorns. The first rains of the season splatter and slide off the hardpan soils. By the end of the season, or often by the end of a single storm, the rain runs off the now-saturated soil. Not all runs off, but enough—one inch of rain over one square mile amounts to 17.38 million gallons of water. The Rillito watershed is almost one thousand square miles. It's not uncommon for two to three inches to fall in a single storm; a single storm once shed ten inches on the Santa Catalina. It rains, it floods.

A RIVER CITY: EXPERIENCE

Tucson as we know it is a strikingly recent phenomenon. In 1900, 7,531 people lived in the City of Tucson. A century later, the city housed 486,699, with the greater metropolitan Tucson area at least twice that number. The desert ecologist Gary Paul Nabhan claims this general "Sun-Belt" population spurt is the single largest human migration in human history. Many came from the East, where rivers run with water and inundate when they flood. Between the 1930s and 1960s, when the population ballooned with 300% annual growth rates and the city sprawled across the alluvial plain, there were no major floods. The rivers barely flowed. Thus when the Rillito and Santa Cruz rivers flooded in 1983, Tucsonans told the *Arizona Daily Star* that the floods were "very strange and frightening"; that "we didn't realize this could happen"; or "I didn't think it could do this." Even the *Daily Star*, having reported Tucson floods since 1879, headlined, "This is a desert?" Afterwards, the Visitors Bureau reassured the citizenry: "The 100-year flood has come and gone, so, by all rights,

Tucsonans should enjoy another century of great Southwest weather." We believed it because it matched our memory: we have been here a short time and we have short lives and shorter memories. Nobody remembers the mesquite *bosques* and beaver dams along the Rillito. There is no time-lapse reel detailing desertification, the slow incision of the riverbed, the water slipping through the sands, the land subsiding beneath our feet.

ANATOMY OF A FLOOD: ENTRENCHMENT

The Rillito, from confluence to confluence, start to finish, is an entrenched system. No more the sliding across the plain, the great leveling and shifting—the river's fixed banks are now wreathed in concrete or armored with rip-rap and spaced wide-enough apart for the big floods to pass. Between the banks is the river. There is one main channel—the channel of the low-volume flows, called the thalweg—and splintering off the main, and rejoining, and splintering off again, are side channels, sought by bigger flows when breaching the overstuffed main. The big, bank-to-bank floods will wipe the palimpsest riverbed clean; lower, more constant flows will rescribe a new thalweg, and midsize flows will reestablish the sundry braids. So confined, the river works its channel. Smaller floods tend to deepen the channel; they'll rub against many of the bends and abrade the banks, but they generally scour the riverbed itself. Large floods tend to widen the river's banks: slamming against the bends, rasping along in helical abrasion. They bore into and burrow under the banks, sweeping away the slumped earth, scooping out the concave meanders in cumulative causation—the greater the curve of the meander, the greater the river rubs against it, accentuating the meander. A big flood, eating at a developed meander, can erode hundreds of feet of earth.

A CITY RIVER: HEADCUTS

In 1889, when the Santa Cruz River still flowed perennially, Sam Hughes, a local merchant and farmer, dug a large irrigation canal that intercepted the river's subsurface flow near downtown Tucson. This was common practice at the time, though most would armor their canals to ensure they wouldn't divert the entire course of the river. Either Hughes failed to do this or the huge flood that ripped through the following year uprooted the armor. It was a flood such as the old-timers reckoned they hadn't seen in twenty-five years. The waters cut into Hughes' canal—which was, again, deeper than the riverbed—and the flood poured off the drop into the canal. Such a waterfall is called a knickpoint—a sudden, often dramatic change in a river or stream's gradient. The flood quickly and consistently eroded the lip of this waterfall, moving the knickpoint upstream—a process called headcutting. The headcutting rippled throughout the river system. Floods no longer fingered through the braids in thin sheets but swerved deep into the headcut, eroding its lip farther upstream. The river below the headcut, focused and powerful, began to incise into an arroyo. With the river running deeper within the earth, the groundwater table dropped to match. The canals that had intercepted the previous surface flow were left, literally, high and dry. The lowering of the Santa Cruz's riverbed formed a knickpoint at the confluence of the Rillito, which slowly migrated upstream, lowering of the bed of the Rillito.

A CITY RIVER: THIRST

The Rillito is not one story, except perhaps that of the climate. The river ran dry, wet, then dry; the river incised, aggraded, then incised again; the river shifted one way then back again, dictated by the ebbs and flows of drought and downpour. In 1858, a local historian described the Rillito as flowing intermittently, lined by cottonwoods, willow, mesquite *bosques* and cattail *cienegas*, broken and stilled by numerous beaver dams. The Rillito floated atop the groundwater table—the shallow aquifer suspended between the Tucson Basin's alluvial fill. The Rillito was, is, the groundwater table—the surface flow only the visible manifest, the tip of the iceberg. In wet spells the aquifer buoyed the surface flows; in dry spells the river shrank into the earth after the diminished aquifer. The Rillito is a story of thirst—in 1891, we replaced irrigation canals with mechanical pumps that tapped the shallow aquifer. Within fifty years we were pumping out more than was being naturally replenished; every year thereafter saw the water table decline substantially. The aquifer dropped thirty-seven feet between 1942 and 1952. The cattail *cienegas* dried and died. The beavers were long gone. The groundwater receded past reach of cottonwood and willow roots, past the mesquite *bosques*, whose taproots plunge eighty feet deep. By 1990, we were sucking 70 billion gallons out of the aquifer a year; in some places, by 2008, the groundwater had dropped 250 feet. Not only do we now have to drill deeper and deeper wells, and not only will we eventually sap the aquifer, but the ground beneath our feet settles into the spaces once saturated by water, the subsidence itself irreversibly affecting the aquifer.

KNOWING A RIVER: THOUGHT

When is a river a river? Or not a river? Is a dry riverbed a river? Is it a river only in the few times a year it flows? Or is that a flood? What if it never flows, if we diverted even the monsoon floods into catchment basins and settling ponds, if the skeleton riverbed never fleshed out, just lay in the sun, year in, year out, would it at some point cease to be a river? We impound the Colorado River behind a dam and call it a lake. We siphon the lake (reservoir? river?) water into an uncovered canal (aqueduct? river?) which snakes across 336 miles of arid desert. The water is pumped 2,900 feet uphill to Tucson, where it is not only injected underground to help recharge the depleted aquifer but mixed with treated sewage effluent and released into the downtown sections of the dry Santa Cruz River in an urban revitalization scheme known as "Rio Nuevo." When is a river a river?

A RIVER CITY: STRUCTURE

Tucson is a grid. Straight-arrow Speedway Boulevard, thirteen miles, east to west. Straight-arrow Kolb Boulevard, nine miles, south to north. Tucson is laid like a tapestry across the alluvial plain, the warp and weft of streets strung between mountains. The city's mechanics, its cybernetic geographies, seem immutable but for the dendrite patterns of arroyos and washes cutting through the circuitry. These urban arroyos have concrete channels and banks; they have natural beds and banks; they are sterile flumes; they are tangled with riparian vegetation, and they lend a certain wonderful, organic disorder to the scene. They lend curves. Sinuosity. Some twine through the city in ribbons of green, then funnel into culverts under perfect square miles of concrete, under clustered mall parking lots, and reemerge farther downstream as though gasping for air. Where the washes still see sky, the roads and blocks of houses press against their contours and mirror their courses as though the city, before the culverts, before the malls, sought to dance, or spoon. Even Kolb, after nine rigid miles, runs against Pantano Wash for a final mile and softens into curves.

A CITY RIVER: LOSS AND ADJUSTMENT

The Rillito is a narrative of loss. Fish, amphibians, crayfish, gone. Thousands of species of plants and animals gone and soon to be gone. To travel to a local, still-extant riparian zone and know an oasis is to know the death of an oasis. The Rillito is a narrative of adaptation. Scraggly bushes attempt riverbed lives between big flows. Raccoons, squirrels, coyotes thrive on the edges the river cuts through the city. Occasionally a mountain lion will slink through, between the mountain massifs, using the dry riverbed as path. Tens of thousands of Mexican Free-tailed Bats, their colonies the largest congregations of mammals in the world, live beneath the Rillito's bridges. Between the rattle and clank of passing cars, the trailing snippets of radio, one can hear the high-pitched squeaking of bats. At sunset the bat chatter rises to a clittering, clackering song. Then, in individual coagulations and communal patches, bats collapse from the cracks in the bridge. They spread their long, translucent wings into flight, not the antic hyper-flittering of some bats, but powerful, swooping loops round the pillars then out into the evening, whirring away towards the Catalina Mountains in great streams, all on the same frequency, all twitching down then up at once, all out into a great bat joygyre that lasts for five, ten minutes and is gone.

ANATOMY OF A FLOOD: CONFLUENCES

The Rillito is a world where single rocks or root systems of buried trees tell the story of the river. Shopping carts obstruct flows for just the few seconds necessary for water to shift between the convoluted braids. Channels ramble and swerve between the slightest variations in slope. Often there is no clue as to the cause of the anastomotic points—the moments of convergence and divergence—as though the channel clove at the resistance of a single grain. Floods are chaotic, surprisingly sensitive, and bristling with variables—velocity, viscosity, volume, pressure, density, temperature—that affect a flood's turbulence, recirculation, eddies, waves. Take sediment loading. A flood interacts with different materials of the streambed differently, and different tributaries supply different materials. Because of its length, Pantano Wash carries relatively fine sediment as it reaches the Rillito, whereas Tanque Verde Wash, flowing right out of the Rincon and Santa Catalina, delivers a relatively coarse load. The mixing of the two flows is like cigarette smoke rising in thin laminae, then suddenly disrupting into turbulence and dissipation. Say a flood has reached a sediment equilibrium—that, fully saturated, it is transporting as much sediment as its variables allow—and in comes a side stream, running clear along bare rock, a different temperature, velocity, and *boom*, the dynamics skew into chaos, into a different flood, perhaps reaching equilibrium again downstream, where *bam*, another tributary disrupts the dynamics. Depending on the variables, different types of erosion will occur simultaneously in varying proportions, overlapping and accentuating each other. Different channels within the main channel will alternately be scoured or deposited. But chaos doesn't signify complete disorder—floods can be chaotic yet exhibit certain general characteristics. Some roll along as semi-consolidated slurries of water, 60 to 80% mud and debris. They'll slide along for miles until some juncture, some critical loss of gradient or speed, when they'll lose their power, and in this exact waning moment the flood loses its peculiar homogeneous cohesiveness and slackens, dumping the heaviest loads. Over the heavy-rock pattern the irregularly pulsing, dying flood deposits lateral terraces, asymmetric ridges and swales, dunes and antidunes, sand patches and scroll patterns.

A CITY RIVER: RUNOFF

Tucson sprawls. A city of parking lots. A city upon which rainwater hits, shards, and runs off. The streets sluice the rain not into storm drains and treatment plants but into the washes, which run into the rivers, which run in flood. The water running off the pavement is different than the water that would run off the desert floor beneath the pavement. What it has gained in car oil, in industrial chemicals, in plastic bags, in dog shit, it has lost in natural sediment. This is important. There was once equilibrium between the sediment scoured from the riverbed and the sediment deposited on a riverbed in runoff flood. No longer. Runoff is free of sediment and running fast and the less sediment it has and the faster it runs, the more it can scour, suspend, and remove without replacing.

A RIVER CITY: STABILIZING INSTABILITY

The Rillito River system is now the story of Tucson. It is a story of control. Take the Broadway Bridge over Pantano Wash. Constructed in 1965, it took but ten years for the Rillito and Pantano's riverbed to lower, or downcut, twelve feet. At that rate the bridge's pilings would have been undermined in two decades. This would not do. In 1979, Pima County enlisted the Army Corps of Engineers to build a $750,000 grade-control structure—essentially a giant concrete ramp—directly downstream of the bridge. The structure is intended to stop headward erosion—any knickpoint migrating upstream won't be able to ascend the ramp to eat at the riverbed around the bridge pilings. Thus the upstream riverbed is maintained at its 1979 level, which, due to continued downstream downcutting, is now some twenty-feet above the downstream riverbed. Or take the Southern Pacific Railroad Bridge over the Rillito, constructed in 1925. The Rillito maintained its perpendicular course under the bridge until a huge flood in December of 1978 ate into 50 acres of land on a western meander bend. This shifted the river channel to the extent that, rather than flowing west, under the bridge, the river now flowed north, adjacent to the bridge. Again, the city sought stability over an instable system. The river still courses to the north, but hits a rip-rapped and soil-cemented bank that pivots the flow into an awkward 90-degree turn, redirecting it back under the bridge. Chaos, controlled.

KNOWING A RIVER: LANGUAGE

If our recent arrival underlies our inchoate understanding of an arid world, our language overlies our ignorance. We betray a linguistic unsteadiness, as though there are objects or processes the English language lacks words for—as it did for saguaro, raccoon, butte, arroyo. "We didn't think it could do this," Tucsonans said of the 1983 flood. "It." The signs along some of the bridges say Rillito River. Other signs say Rillito Creek. Pima County's Regional Flood Control Department website alternates between creek and river. A stream can be a brook, beck, burn, bourne, nant, creek, crick, kill, lick, rill, river syke, bayou, rivulet, or run. A tributary can be a branch, fork, or prong. A dry or ephemeral river can be an arroyo, wash, draw, wadi. Yet the Rillito, because it is dry, or short, or ignored, is baffling as creek or river or "it." The same when in flood: "I didn't realize this could happen." This. A flood, generally, is when a river exceeds its channel's capacity and water inundates the surrounding floodplain. Flood is inundation, and if 20,000 cubic feet of water per second (c.f.s.) is roaring down the Rillito, gouging out hundreds of feet of bank and destroying millions of dollars worth of property, yet not breaching the banks, not inundating, many will still hesitate to call it a flood. To hydrologists, engineers, it is a "significant flow event" with perhaps some "corresponding lateral erosion." Some call it a flood. On one level, this is mere semantics, needless quibbling; on another, our linguistic constructions define our meanings. We name the world as we know it, or, perhaps, as we don't know it. Our understanding remains in suspension; our vocabulary has yet to put down roots.

KNOWING A RIVER: DISASSOCIATIONS

We don't know the Rillito as a river because we don't need to know the Rillito. It has little or no bearing on our day-to-day existence—it flows but a few times a year and destructively maybe once in a lifetime. We turn the tap and out flows clear water. We are equally removed from water's origins if it comes out of a plastic bottle or from hundreds of miles away or thousands of feet underground or from hundreds of miles away to be injected thousands of feet underground. We don't know the Rillito because it's not water—it's dust, and rock, and scrabbly vegetation, and trash, and when it's water it's more sludge than water. The connections are not made. This is not a judgment; this is the world as we know it—where disassociations reveal more about ourselves than associations.

A RIVER CITY: FLOOD

Tucson was reminded of its rivers in 1983. It had been the second wettest September on record when Tropical Storm Octave delivered five days of steady rain, culminating in a 24-hour, 3.58-inch deluge on the third of October. The Rillito roared at 29,700 c.f.s., the Santa Cruz a staggering 52,700 c.f.s. For three days, Tucson's rivers dwarfed the Snake or the Hudson or the Colorado. Trees, telephone poles, cars, sections of bridges, mobile homes, and culverts swept along in the flood. Landfills became lakes. Natural gas lines exploded. Sewage lines ruptured, adding thousands of gallons of raw sewage to the burgeoning flow. People stuck in river-crossings were swept away. People rescuing people stuck in river-crossings were swept away. People watching the rising floodwaters were swept away when the undermined banks beneath them collapsed. The flood killed twelve people, destroyed all but one of the bridges across the Santa Cruz, and caused 250 million dollars in damages. Out of the receding floodwaters emerged floodplain ordinances. Future riverbank development was required to be set 500 feet from the existing banks of the Rillito and Pantano unless approved by the county's Chief Engineer. Developers were allowed to ignore setbacks if they stabilized the banks. Consequently, stabilization became the primary solution to flood management. We fixed the meanders in place or straightened the river's sinuosity. We widened the channel. We sprayed the majority of the Rillito's banks with soil-cement or overlaid them in rip-rap or fencing. We kept building.

A CITY RIVER: ACCEPTANCE

The Rillito is a narrative of change. We can't go back to the days when Tucson didn't edge against the Rillito. We can't go back to when the Rillito's mouth stretched a mile wide or its course shifted by miles in single floods. The Rillito is strung between golf courses and cottonwoods, development and open space, bats and bridges, and ideas about how it is and how it was. The Rillito will not flow again but in flood. The aquifer will not rebound. We can drop Tucson's per-capita water consumption from its current 177 gallons a day to 100, we can xeriscape every yard, use reclaimed water for all nonpotable necessities, cut agricultural and industrial uses of groundwater—and, with current population growth, still overdraw the aquifer. No amount of conservation or effluent use or injection of canal water can match our population growth. Based on current weather patterns—without even taking into account that it's projected only to get hotter—Tucson's water use is projected to double in the next fifty years.

A CITY RIVER: FUTURE

The problem is, piecemeal flood protection doesn't work. A smooth concrete riverbank only amplifies a flood's velocity, providing it that much more power to bore into the next bend's unprotected bank. If upstream protection begets downstream protection, at some point the entire river's banks will have to be systematically stabilized. Not to mention the riverbed itself. The huge flood of 2006, while not overtopping banks or gouging out meanders, lowered Pantano's bed beneath Speedway Boulevard eight feet. This type of downcutting threatens to undermine even where stabilized banks or grade structures toe-in with the natural channel—a knickpoint cannot migrate up and over the concrete check-dams, but it can corrode the riverbed from their bases. We armored the confluence with the Santa Cruz, so any downcutting the Santa Cruz incurs is not transferred upward. But sediment-deficient runoff will still scour the Rillito's riverbed. The future of the Rillito is that of a continuous, tightly stepped series of grade control structures, similar to rice terraces, or we'll simply line the riverbed with concrete and the Rillito River will become a flume.

CONFLUENCES: CITY AND RIVER

The story of the Rillito isn't just the overdrafting of the aquifer, the artificial headcuts, the impermeable streets and runoff, the inadvertent yet profound morphological changes, but every point, every confluence, every influence. The seemingly unrelated factors that tie in everywhere—buffelgrass replacing native vegetation so that the Sonoran desert is primed for a massive, unprecedented burn and subsequent unprecedented runoff. A cow browsing a desert shrub, the withered roots no longer trapping sediment, no longer absorbing water, the next storm's rain thus running onto the path the cows trod, making it a rill, eventually an arroyo. Gravel and sand is mined out of the riverbed to build roads that block natural sediment flows. The mine pits act as giant sediment receptacles, allowing flood waters to momentarily pause, deposit their sediment loads, and then scour more from the banks and riverbed downstream. Or the pits act as scour holes, with headward erosion gnawing upstream out of them. The Rillito is all these stories. But the bats, too, tell the story. Bats and the biomimicry of streets to washes. Bats and the rare, sudden, and human awe of a dry river in flood.

Rillito River Project

ELLEN BENJOYA SKOTHEIM Creative Director
JOHN NEWMAN Technical Director
JENNIFER POWERS-MURPHY Publicist

The Rillito River Project fuses art and science to create ephemeral environmental installations and performance art events on a dry riverbed site in Arizona. Inaugurated in 2007 by an international group of artists, architects, musicians, writers, builders, and anthropologists, the goal of the Rillito River Project is to raise awareness of the devastating effects of global climate change on the vanishing rivers of the Southwest. One of the Project's first public events, presented in October 2007, was the world premiere performance of New York Metropolitan Opera violinist Patmore Lewis' original composition, "Elemental Flow." This piece tells a story of the history of the Rillito River, a story as old as creation and as current as today.

One of the Project's key endeavors is the annual Bat Night event, which serves to increase consciousness of the delicate balance of life in river ecosystems in the desert. At the Campbell Avenue Bridge over the Rillito, every evening from April through October, approximately 45,000 Mexican Free-tailed Bats exit the bridge's expansion joints that serve as their migratory home and take a dramatic nightly flight in search of food. In honor of this natural event at a major crossroads in the city, Bat Night has been held every year since 2008, drawing thousands of people into the dry riverbed to witness the flight of the bridge's transient residents, and to experience various artistic presentations that not only entertain but also educate about the environmental issues facing the Southwest.

The first annual Bat Night with general public participation was held on September 13, 2008. More than 2,000 people gathered near sunset in the Rillito riverbed east of the Campbell Avenue Bridge to watch the bats fly out from their home. An hour prior to this natural phenomenon, Ellen Skotheim, Rillito River Project Creative Director, opened the event with a reading of "The Bat," a poem by Theodore Roethke, and a (water) toast to the soul of the river. Bat expert Yar Petryszyn of The University of Arizona presented a lecture on bats, contrasting the dark and mysterious portrayal of bats in folklore and mythology with the facts, such as bats' important role in plant pollination, and their voracious appetite for insects.

For Bat Night 2009, the Rillito River Project presented a sky chart made with seven weather balloons, each eight feet in diameter, representing a mirror-image water table. The balloons were raised into the sky above the dry Rillito riverbed to demonstrate the water table's dramatic plunge of 170 feet between 1945 and 2005, a drop resulting from increased groundwater pumping to service a population that grew from 40,000 to one million residents, coupled with a multiyear drought.

Bat Night 2010, incorporating design, architecture, and literary projects created by University of Arizona students and faculty, was held on September 11, and reflected on loss. In 2012 the Rillito River Project collaborated with Barbea Williams and the Rio Vista Elementary School on Welcome Back the Bats.

Bat Night is only one of the programs created by the Rillito River Project. In 2010, the Project established the Border Biosphere Art Lab, which brings together nationally and internationally acclaimed artists to meet with University of Arizona scientists and environmentalists to develop their individual artistic response to climate change. Both Bat Night and Border Biosphere Art Lab involve close collaboration with The University of Arizona.

OPPOSITE PAGE TOP Fishing on the Rillito during Concept Week, 2007 | photo: Vidar Lerum

OPPOSITE BOTTOM Patmore Lewis performing "Elemental Flow," 2007 | photo: Vidar Lerum

LEFT Flam Chen gearing up for Bat Night 2009 | photo: Victor Silverman

ABOVE Ellen Skotheim's water toast at Bat Night 2008 | photo: Kimble Cohn

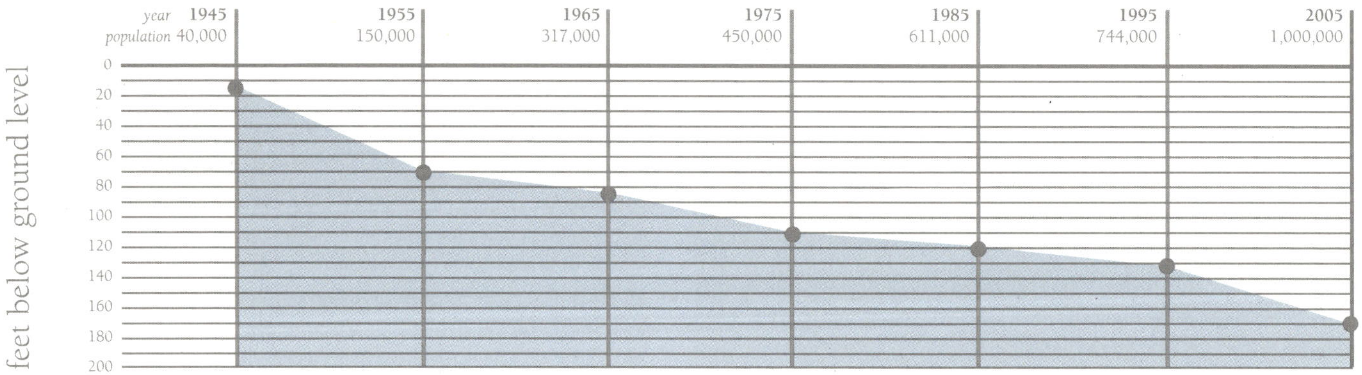

TOP *Water Table* | authored by Chris Tanz and performed by Flam Chen | photo: Tim Fuller

BOTTOM Groundwater graph | design: Kelly Leslie | 2010

OPPOSITE PAGE TOP LEFT Sand pail in use at Bat Night 2010 | photo: Jonathan Black

OPPOSITE PAGE TOP MIDDLE Participants at Bat Night 2010 | both photos: Tim Fuller

OPPOSITE PAGE TOP RIGHT Bat Night 2010 poster | screen print | design: Jonathan Black

OPPOSITE PAGE BOTTOM Bat Night 2010 | photo: Tim Fuller

BAT
NIGHT
2010

Bats Sleeping

BAT
NIGHT
2010
THE FUTURE UNFOLDS IN A SANDBOX
A RILLITO RIVER PROJECT EVENT
SATURDAY, SEPTEMBER 11, 5:30PM
BAT NIGHT 2010 DONORS

TOP **Bat Dance performed by Rio Vista Elementary School children and Barbea Williams Dance Company | photo: Tim Fuller**

BOTTOM LEFT ***Welcome Back the Bats*** **| archival inkjet art print | 9" x 6" | design and illustration: Robert Long**

BOTTOM RIGHT **Welcome Back the Bats 2012 | photo: Tim Fuller**

waterBar

NICOLE SWEENEY

B Arch, College of Architecture + Landscape Architecture, 2010

Capstone Design Studio 2010

Every year millions of gallons of water are channelled away from the Rillito and its landscape as a result of the concrete embankments that contain it. Consequently, the Rillito has shifted from a natural to an artificial place where infrastructure compromises the primary function and performance of the river. The goal of this project was to investigate the relationship between the natural course of the Rillito and the artificial technologies that have been imposed on the river and its surroundings. The proposed installation, *waterBar*, responds to the dichotomy between the natural and artificial landscape.

The proposed installation is composed of 237 waterBars placed on a grid within the bed of the Rillito. Each waterBar is constructed with 80 water bottles, representing the volume of water lost over the course of one year in one square foot of the Rillito due to the hard infrastructure. The waterBars align on north–south and east–west axes to reinforce the dichotomy between the natural landscape and the Cartesian grid structure that retains its ordered form regardless of the local conditions. The use of the generic water bottle as an installation material is intended to bring awareness to the artificial containment of water at local and global scales.

The installation's composition aims to illustrate both the natural and artificial components of water consumption and containment. Through the use of artificial materials and the imposition of a grid on the river landscape, the project seeks to alter the perspectives of local residents who interact with the Rillito on a daily basis.

waterBar | water bottles, etched plexiglass, threaded rods | 143″ x 12″ x 12″ | 2010

Confluence of Boundaries

JENNIFER HEINFELD

B Arch, College of Architecture + Landscape Architecture, 2010

Capstone Design Studio, 2010

This project began with an analysis of the place and the people of Tucson. With its diverse demography, political contention, and immeasurable landscapes, this desert city offered vast research possibilities. Most compelling was the contemporary conflict, found 45 minutes away, at the Arizona–Mexico border. Historical research, mappings, and census analysis revealed the relevance of this conflict to our own community and its presence within Tucson's borders.

The proposal intends to expose the richness of the site—a waterless, abandoned arroyo, at places 300 feet in width, segregating a community by the scar it imposes. In response to the Arizona–Mexico border, the specific project site gained potency through its many existing boundaries—the clear natural demarcation, the few pedestrian bridges spanning the river, and the invisible dividing line of the city limits. The proposed design interprets these lines, taking form as a site-specific spatial installation in the bed of the dry Rillito, to make visible the often-invisible boundaries within our landscapes—to expose the hard political boundary, represented by the city limits line, and the soft ambiguous boundary, related to the natural conditions of the site.

Using the existing riverbed soil and sandbags as the installation medium, selected for their symbolic and ephemeral nature, the research would be interpreted as a series of walls that map these divisions, adapted in their patterning, height, and width based on material availability and programmatic needs. The spaces created by and within these walls challenge the existing boundaries and bring awareness to their dividing impact. Acting as a mediator between the north and south banks, the river becomes a place of gathering, performance, collaboration, and social tolerance, where nature takes precedence over the arbitrary limits of a city.

Beyond research and design, prototyping was done in three installments; this was recorded and turned into a short film documenting the testing of material. By critically analyzing a region, a site, and the demographics of a place, the project intention is to promote public understanding and acknowledgement of the borders within our own communities.

For Bat Night 2010, this project was adapted to create a stage, public seating, and an area in which children played in the sand.

OPPOSITE Mapping Tucson: (1) city limits; (2) street patterning; (3) flora and fauna in relation to elevation/biome; (4) population density in relation to ethnicity; (5) parcel size; (6) public/private amenities.

1

2

3

4

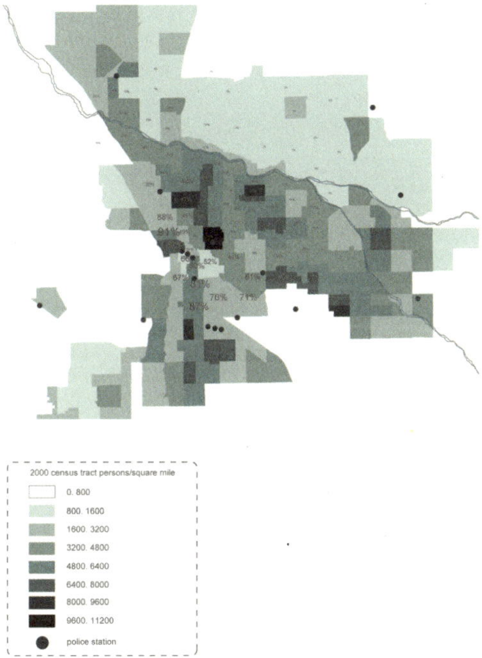
2000 census tract persons/square mile
0. 800
800. 1600
1600. 3200
3200. 4800
4800. 6400
6400. 8000
8000. 9600
9600. 11200
police station

5

6

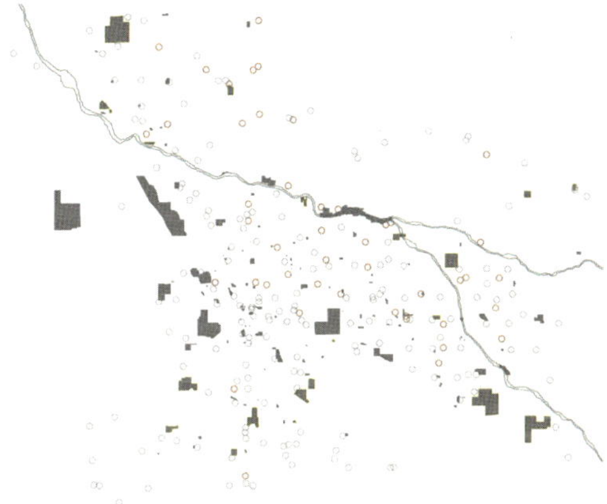

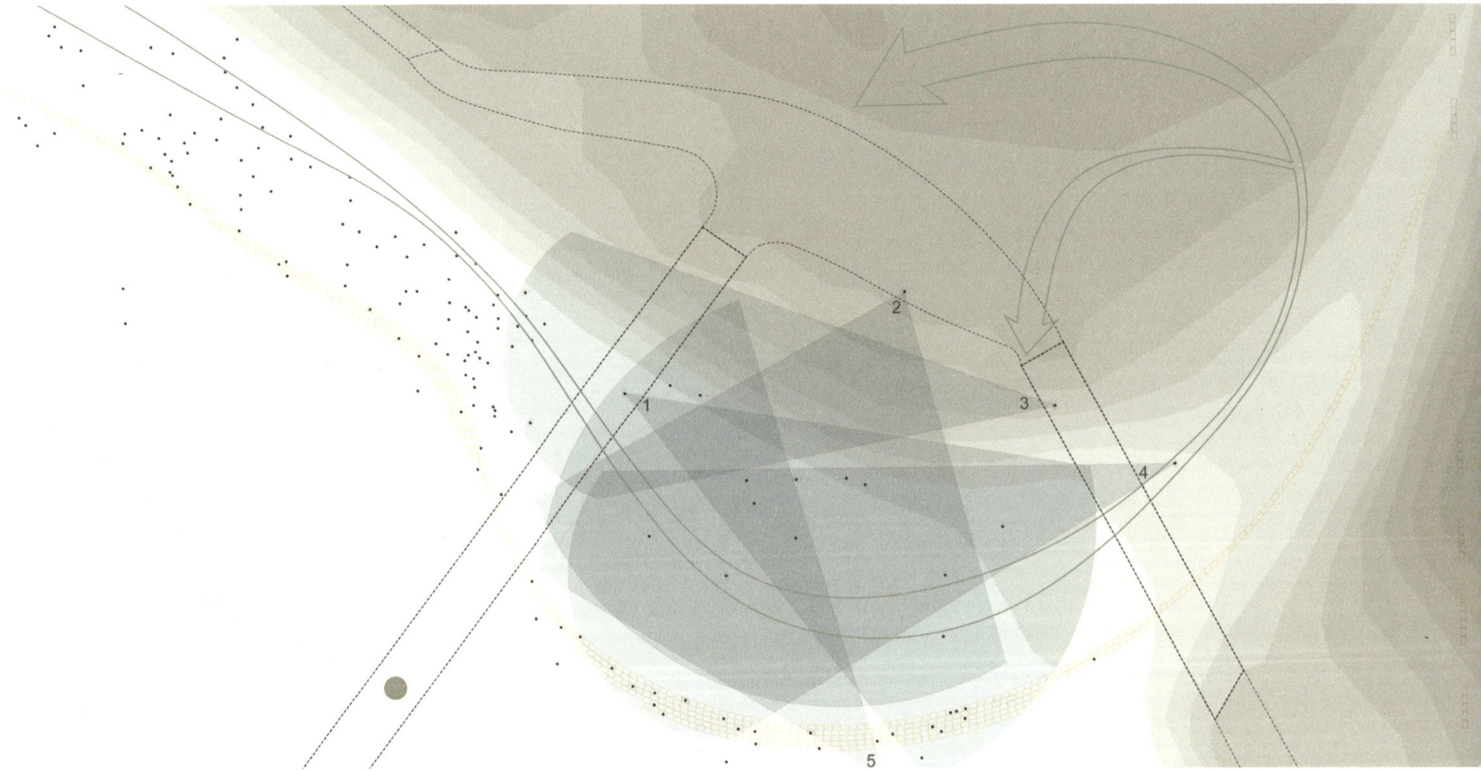

ABOVE TOP Top Row: Construction studies for the political boundary | Middle Row: Construction studies for the social boundary | Bottom Row: Ergonomic studies and human interaction.

ABOVE BOTTOM Detailed site plan of the amphitheatre, showing manifestation of the social and political boundaries, anticipated density of spectators, and cones of vision relating to perspectives.

ABOVE TOP View looking west.

ABOVE BOTTOM LEFT TO RIGHT View looking east from bridge over the Rillito; view from riverbed at north bank; view looking south towards bridge; view looking west into the riverbed.

What's So Important About Water?

JESSICA GERLACH

MFA, Visual Communications, School of Art, 2011

Independent Study and MFA Thesis, 2010 and 2011

ABOVE Illustration details from Bat Night 2010 banners | inkjet on vinyl | 6′ x 3′ | 2010

OPPOSITE PAGE Banners installed for Bat Night 2010 | inkjet on vinyl | 6′ x 3′ | 2010

As a designer and illustrator, I am interested in the possibilities of visual communications as a means of creating a visual dialogue about issues of social, political, and environmental relevance. My work is fueled by my commitment to use art and design to inform, educate, and engage the community. My thesis project, *What's So Important About Water?*, employs visual communications, illustration, and graphic design to create an interactive educational experience for community members that emphasizes water and ecology, and their significance to the species that once inhabited the Rillito.

The outreach project was designed to build on the goals of the Rillito River Project, an international group of artists, architects, musicians, writers, builders, and anthropologists who encourage and support a series of ongoing events in which artists conceive of and plan artwork to be installed or performed in the Rillito riverbed. These ephemeral environmental installations and performance art events bring people to this site to raise awareness of the effects of global climate change on the vanishing rivers of the Southwest.

The goals of my project were to increase public awareness about the importance of water conservation for the ecology of the Rillito Watershed, and to frame water education in a way that emphasizes the importance of native plants and animals as indicators of a healthy ecosystem. To achieve these goals, the project created connections between science, design, and the community, and employed extensive interdisciplinary research and collaboration.

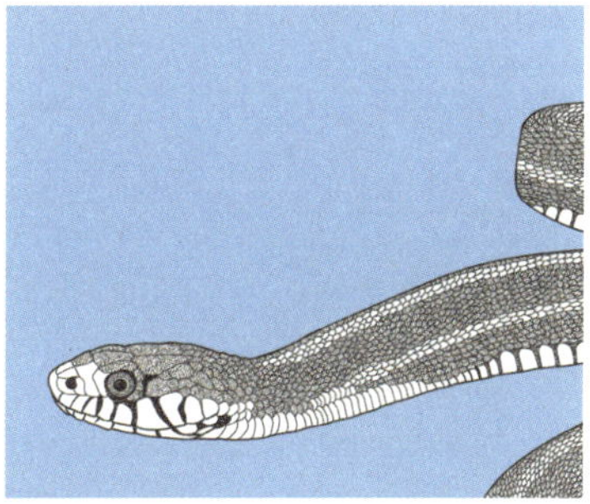

I began the project by researching and consulting with environmental experts to determine eight key species that once thrived in the riparian habitat of this Southwestern river. These species would serve as ambassadors to the lost river habitat. They were selected based on their importance to the ecology of the river and their classification as endangered, threatened, or "a species of concern." For example, the Fremont Cottonwood is considered a keystone species for riparian areas and has a direct connection to the health and survival of other species, including the Western Yellow-billed Cuckoo. I created illustrations of each of these species and incorporated them into banners and take-away collectable cards to be used as key components of Bat Night 2010, held on September 11.

For the event, I enlisted the help of science experts, who spoke with audience members and distributed the illustrated cards. The six-by-three-foot banners served as the visual keystone for the installation, and, along with the orange sandbags, helped to delineate a play area in the river for children and families attending the event. The resulting public environmental installation informed an audience of 5,000 people about key plant and animal species that have been lost with the deterioration of the riparian habitat in one of Tucson's rivers.

After the event, I continued to research and develop artworks about these key species. I wrote and illustrated six artist books on the native animals highlighted at the event: the Gila Topminnow, the Giant Spotted Whiptail, the Lowland Leopard Frog, Merriam's Mesquite Mouse, the Northern Mexican Gartersnake, and the Western Yellow-billed Cuckoo. These books were designed to tell the stories given by the science experts about each of these species.

ABOVE TOP **Bat Night 2010**

ABOVE BOTTOM LEFT TO RIGHT ***Western Yellow-billed Cuckoo*** | **inkjet** | **7.75" x 8"** | **2011:** ***Giant Spotted Whiptail*** | **inkjet** | **7.75" x 8"** | **2011**

OPPOSITE PAGE **MFA Exhibition installation** | **The University of Arizona Museum of Art** | **2011**

What's So Important About Water?
What's so important about water? is an artwork outreach project designed to build on the goals of The Rillito River Project, an international group of artists, architects, musicians, writers, builders, and anthropologists, who encourage and support a series of ongoing events in which artists conceive and plan artwork to be installed or performed in the Rillito riverbed in Tucson, Arizona. What's so important about water? uses visual communication, illustration, and graphic design to create a series of banners, take-away collectable cards, and an interactive educational experience for community members that emphasizes water and ecology, in addition to its significance for the species that once inhabited the Rillito River. The banners and the interactive experience were a component of "Bat Night 2010" held on Saturday, September 11, 2010.

Canto Rillito (With Bats Under Bridges)

LOGAN PHILLIPS

El bisabuelo de tu tatarabuelo es ahora
aquel arroyo que serpentea en la montaña.
— Eduardo Galeano

Rillito rillín, riote río
Rio-ito más pequeño cada año
Rillito rillín, riote río
Cada año, más arena que agua

Bats get a good day's sleep
while the city rushes in circles.
Bats whose new caves are under the bridges
built to hold the hurry of the city:
overpass caverns.

Do bats dream in sonar, wings pulled tight?
What is the pitch of high noon?
What does a stoplight sound like?
Who are the oldest poblanos of the Old Pueblo?

Rillito rillín, luego cae la noche, radiante.

At dusk, you can watch the shadows flip over themselves
in somersault as the sun sinks its heat
on the heavy mirage of the horizon jagged.

The saguaro sundials stand still.
The sunset stops traffic,
El atardecer arde prendido.

This is when the desert inhales
after holding its breath all day, the moment
the bridges let their cement bones creak
when the skies let their colors leak.

This is when the desert lets its life show,
ghost sand water begins to flow,
life comes out into the night.

Night flowers begin
to bloom, a reflection of stars.

Murciélagos, estos residentes persistentes,
esta gente desértica, voladora, bailadora
dip duck dive feeding through the sunset:

bats, bits
of night
sky broken
off, flirting
with fading daylight,
a living cloud of ink.

¿De qué son los sueños de murciélagos?
¿De bichos deliciosos y voladores?
¿De un Rillito con agua?
¿De las noches sin faroles?

Río-ito, agua mito
aun en agosto.

Rillito we call you río-ito
little river smaller every year,
dimunitivo.

After every year,
Rillito rechiquito.

¿Pero es río pequeño
o es río empequeñicido?
¿Quién traga el Rillito enterito?
What are the bats dreaming about,
as their wings bend in ballet?

Who gets invited to eat at the water table?
Who has lost their water table manners?
Who drinks too much?
Who remembers the Rillito as more than sand?
Who drinks eight glasses a day?
Who putts eighteen holes a day?
Who sprays pesticides every day?
Who drinks too much, leaving only sand?

Flirting with the first dark of night,

esta gente desértica, voladora, bailadora,

the living night of this desert,
the living desert in this night,
murciélagos sobrevivientes
floreciendo bajo los puentes.

Murciélagos sobrevivientes
floreciendo bajo los puentes.

Desert Water: Paradoxes and Trade-Offs

GREGG GARFIN

The key climate, environment, and resource management issues in the Southwest all revolve around water. From the most remote outpost, seep, or spring to the largest farm, city, or river in the region, the quantity and quality of water is fundamental to the well-being of people, wildlife, ecosystems, rangelands, everything we value. This is true everywhere on earth, but it comes into sharp contrast in the arid southwestern United States and northern Mexico, home to four deserts (Chihuahuan, Sonoran, Mohave, Great Basin), where surface water is scarce, perennial streams are rare, most lakes are man-made, and human population growth and economic activities have increased at rapid rates. This brief essay explores several Arizona water issues, including climate and societal challenges to sustaining water resources in this urbanizing, arid region. The essay is not meant to be comprehensive, but illustrative of the science, history, paradoxes, and trade-offs behind water-related issues addressed by the projects in *Ground/Water*.

REGIONAL WATER ISSUES

The Colorado River, the main source of surface water for southern Arizona, has been called "the hardest working river" (Pulwarty et al. 2005). By the standards of other major continental rivers, the Colorado is quite small. It is about 2.5 times shorter than the Mississippi–Missouri system, and only around one-third the length of the Nile. Its discharge is only one-tenth that of the Nile, one-hundredth that of the Mississippi–Missouri system, and one-thousandth that of the Amazon. The Pacific Northwest's Columbia River, which drains approximately the same area as the Colorado, discharges nearly 40 times the flow of the Colorado at its mouth. Nonetheless, the Colorado is an engine of economic growth and sustainability for major cities such as Las Vegas, Los Angeles, Phoenix, San Diego, and Tucson. For cities such as Phoenix and Tucson, which have depended on groundwater supplies for most of their history, water from the Colorado River is referred to as a renewable water supply.

The Colorado River is supplied primarily by snowmelt from headwater mountain ranges in Wyoming, Utah, and Colorado; 15% of the land surface generates 80% of Colorado River runoff. Snowmelt from these "water towers" of the Southwest feeds an intricate system of dams and diversions that delivers water to farms and urban areas hundreds of miles from the main stem of the river. In the twentieth-century era of dam building in the western United States, the U.S. Bureau of Reclamation developed substantial storage in Lake Powell and Lake Mead which, together, can store up to four years' worth of Colorado River water supply. These massive reservoirs, the largest man-made lakes in the United States, feed the temporary holding "ponds" of Lake Mohave and Lake Havasu, which are the points of conveyance for water to major downstream agricultural areas, such as the Imperial Valley, and urban areas like Los Angeles, Phoenix, San Diego, and Tucson. To reach Tucson, Colorado River water is lifted over 2,900 feet from Lake Havasu and channeled over 300 miles through the Central Arizona Project canal.

Of course, the Southwest is a region prone to historic multiyear droughts and prehistoric multidecade "megadroughts" (Meko et al. 2007), the likes of which have not been experienced by our technologically advanced, highly urbanized society. A drought which began in the late 1990s rapidly reduced water storage in the great reservoirs of the Colorado River, and put the specter of water shortage, and the prospect of climate change, in the consciousness of many regional water planners and decision makers. The ongoing drought has also brought home to the public the well-documented fact that the 1922 Colorado River Compact doled out more water to the seven basin states than is available on average, because the data available in 1922 characterized as average what is now known, with a much longer streamflow record, to have been the wettest period on record. Briefly, the River's water resources are overallocated. Concerns over potentially scarce resources and the prospect of shortage, along with a history of Arizona–California legal battles, brought this point home to the state of California, which was forced by the Department of Interior, in 2003, to reduce its historic overuse of Colorado River water supplies. To keep from losing water to California, Arizona has since the 1990s banked Colorado River water underground. This use of surface water, illustrative of the "use it or lose it" mandate of western water law, is one of the many conundrums of the water world explored in *Ground/Water*. Paradoxically, water banking is also a practice advocated to resist potential impacts of future climate change, as it provides a hedge against future shortage.

Another conundrum is that more than 70% of Colorado River water goes to agriculture, an activity considered a "beneficial use," which is a cornerstone of western water law. Desert agriculture is remarkably productive, due to the ability to grow multiple plantings during a single year. However, for rivers such as the Colorado, mounting pressure has been put on surface water supplies from drought, increasing population growth, and demands to reserve or return water to restore and preserve environmental processes. Add climate changes, such as hotter temperatures and increased evapotranspiration, and it is not difficult to envision a future in which there will not be sufficient water for everything to which we have grown accustomed. Society will need to make tough decisions, explicit trade-offs, about the values and uses of water. Paradoxically, the looming crisis of insufficient surface water supplies from the hardest-working river has fostered much more public input into debate and increasingly creative decision making about future management practices, planning, and preparations.

TUCSON: WATER, WATER, EVERYWHERE? ANYWHERE?

One way in which the contributors to *Ground/Water* focused their explorations on the theme of water was to examine the specific situation of Tucson, Arizona. Like other Arizona cities and towns, Tucson has a long history of human habitation, a rapidly increasing population, and an enormous and well-structured history of water development and management accompanied by unfortunate, some would say catastrophic, unintended consequences for surface water and ecosystems. Tucson's water challenges are similar to those throughout the Southwest. History and hydrology show that the cumulative impact of Anglo-American settlement has contributed to the overutilization of surface water resources, the dewatering of streams, aquifer decline, land subsidence, and channel erosion. By 1940, groundwater was being withdrawn at rates greater than natural replenishment (City of Tucson Water Department 2004). By the 1970s, it was recognized that the rapid rate of groundwater pumping, initially by agriculture and later by urban development, threatened groundwater supplies, infrastructure (through subsidence), and water quality.

In 1980, Arizona passed progressive groundwater management legislation that regulates further agricultural development, and well drilling, and requires developers to prove a 100-year water supply for new housing subdivisions. However, these safety valves only apply to around 13% of the area of the state—the so-called Active Management Areas (AMAs), of which Tucson is one.

Loopholes in both water and development laws allow (a) "wildcat subdivisions" in which large parcels are subdivided in a way that allows the developer to avoid providing some services (e.g., sewer lines) and to use unmonitored groundwater wells, and (b) chicanery, such as skirting the state's assured water supply law (outside of the AMAs only). With the latter, if a parcel is found to have a less than 100-year water supply (i.e., "inadequate water supply"), the developer can proceed with developing the lot, as long as they disclose the inadequate supply finding to the initial sale (Jacobs and Stitzer 2007). If the land changes hands, the seller is not obligated to inform the next buyer of the inadequate supply. State agencies can only issue the initial findings, and have no enforcement capability beyond that. Thus, we have another paradox, this time acknowledgment of a bad situation—unsustainably overpumping groundwater—led to progressive legislation, which has been undermined (to some extent) by legal loopholes embedded in the legislation.

> The case of the Colorado River Delta is perhaps the most dramatic example of the trade-offs that have been made between economic and environmental values.

DEWATERING OF LOCAL RIVERS AND STREAMS

Some of the details of Arizona water law were astonishing to the *Ground/Water* participants, even though they may make perfect sense to the legislators and judges who have crafted and interpreted them. When describing these details, one is often left with the feeling that the emperor has no clothes. These facts are in plain sight in the public record, and yet are obscure to many Arizonans. For example, there is the well-known disconnect between surface and groundwater in Arizona water law. Arizona law distinguishes between surface water, subflow (a concept, developed in the late nineteenth century, to describe water moving through underground channels in a way that is connected to surface water), and groundwater, which percolates rather than flows (Glennon 2007; Pearce 2007). The science of hydrology makes no such distinction, noting the important connection between surface and groundwater, but the subflow concept has persisted in water law, to the detriment of substantial surface streams. For example, in the San Pedro River basin southeast of Tucson, new wells drilled in the expanding city of Sierra Vista are capturing water moving underground from the Huachuca Mountains toward the river. Such water, which is not characterized as subflow, can be pumped without restriction, ultimately robbing stretches of the river of water that is much needed to maintain stream continuity during the drought episodes to which the region is prone (Glennon 2007).

THE AGE AND ABUNDANCE OF GROUNDWATER

It is easy to take for granted the abundant groundwater that flows from our faucets and has allowed for the development of large urban settlements and commerce in the desert Southwest. While water is present under the land surface in all areas, it can only be withdrawn in amounts sufficient for use by humans where the hydrology and geology are favorable. Isotope dating studies by USGS researchers show that groundwater in the Tucson Basin is several thousand years old (Bexfield et al. 2011).[1] In only a brief century or so, we have depleted or "mined" a substantial amount of this slow-to-renew resource. This makes groundwater a particularly precious commodity, to be used sparingly, or as a backup supply for sustained droughts. While Arizona has vast quantities of groundwater in storage, depletion of groundwater can result in undesirable local consequences, such as land subsidence (over 20 feet in some parts of Arizona) and capture of flow from nearby streams and springs.

Like every other western state in the United States, Arizona uses the doctrine of prior appropriation, or "first in time, first in right" to govern surface water diversions. This law "encourages economic development at the expense of treating all activities as equal in value," and thus creates perverse incentives to undertake intensive water diversions (Glennon 2007). The dewatering of streams, which, in the 1800s, seemed like a reasonable price to pay for the economic development of the vast West, is a consequence of this practice. Early Arizona lawmakers also adopted a "reasonable use" doctrine to govern groundwater allocations. Reasonable use allowed unlimited pumping of groundwater as long as the water was put to beneficial use on the land, which, in Robert Glennon's words, "encompasses almost all pumping." As Glennon mentions, these concepts may once have been useful, and the ramifications of developing new water laws could surely be a legal nightmare, but this inertia in the legal system leads to the stark reality of the desiccated riverbeds surrounding Tucson.

> Daily we flush the Colorado River down our toilets, or without even a passing nod to the snow-covered headwaters in Wyoming, we pour the Colorado onto our gardens.

THE COLORADO RIVER DELTA

The case of the Colorado River Delta is perhaps the largest and most dramatic example of the trade-offs that have been made between economic and environmental values. Typically, as rivers flow downstream and acquire greater volumes of input from tributary streams, they discharge their greatest volume at their mouths. These mouths often fan out in wide deltas and rich estuarine environments, where freshwater meets ocean water. In the Colorado River system, the history of upstream diversion and legal agreements made initially on the U.S. side of the border treat flow to Mexico, and hence the Colorado River Delta, as an afterthought. Of course, Mexico, too, has opted to value terrestrial economic activity over riparian and estuarine environments by diverting most of its paltry Colorado River allocation for agricultural use in northern Baja California. This has resulted in a situation whereby the river only rarely flows all the way to the ocean, petering out in a now-dry delta littered with the calcic carcasses of the shellfish and marine life that once thrived there.

RIPARIAN AREAS

The Colorado River Delta is a large-scale example of a phenomenon repeated in many rivers throughout the arid Southwest: the desiccation of streambeds and riparian habitat. The local example studied by several of the participants in *Ground|Water* is the dry bed of the Rillito, or "little river." The Rillito, which drains the Santa Catalina and Rincon mountain ranges, seldom flows, except during exceptional thunderstorms or relatively rare multiday rain events. As noted elsewhere in this book, the Rillito has suffered numerous local extirpations of wildlife species, and the loss of much of the cottonwood, sycamore, and mesquite gallery forests and bosques that once lined its banks. The stream dewatering is the result of groundwater mining, which has reduced the stream's baseflow level to beneath the surface. The bulk of riparian forest loss is due to a combination of groundwater drawdown and development priorities in which streamside forests have no place.

The case of the Rillito, studied closely by *Ground|Water*'s participants, poignantly illustrates the paradoxes and trade-offs associated with desert oasis cities. On the one hand, society, perhaps unwittingly, traded the rich aesthetic and ecological miracle of a perennial stream in the desert for extensive irrigated agriculture, initially, and the economic and social opportunities of a winter haven metropolis, eventually. As part of this trade-off, Tucsonans contribute to the ongoing dewatering of the Delta and the robbery of riparian flows in parts of the Colorado River. Daily we flush the Colorado River down our toilets, or without even a passing nod to the snow-covered headwaters in Wyoming, we pour the Colorado out onto our garden beds. Even the virtuous community gardener eats the Colorado River, or the thousand-year old "fossil" groundwater, in every bite of locally grown lettuce. On the other hand, over one million people and a substantial local commercial economy could not exist here without doing substantially more environmental damage to local ecosystems and aquifers, were it not for the water that we import from the Colorado River.

NOTES

1 Groundwater along stream channels is likely to be substantially younger than groundwater in the central part of the basin.

REFERENCES

Bexfield, Laura M., Susan A. Thiros, David W. Anning, Jena M. Huntington, and Tim S. McKinney. 2011. "Effects of natural and human factors on groundwater quality of basin-fill aquifers in the southwestern United States—conceptual models for selected contaminants." *U.S. Geological Survey Scientific Investigations Report 2011–5020.*

City of Tucson Water Department. 2004. "Water Plan: 2000–2050." Accessed March 18, 2012. http://cms3.tucsonaz.gov/water/waterplan and http://cms3.tucsonaz.gov/water/waterplan-2008.

Glennon, Robert. 2007. "The disconnect between water law and hydrology." In *Arizona Water Policy: Management Innovations in an Urbanizing, Arid Region,* edited by Bonnie G. Colby and Katharine L. Jacobs. Washington, D.C.: Resources for the Future Press.

Jacobs, Katharine L., and Linda S. Stitzer. 2007. "Water supply and management in rural Arizona." In *Arizona Water Policy: Management Innovations in an Urbanizing, Arid Region*, edited by Bonnie G. Colby and Katharine L. Jacobs. Washington, D.C.: Resources for the Future Press.

Meko, David M., Connie A. Woodhouse, Christopher A. Baisan, Troy Knight, Jeffrey J. Lukas, Malcolm K. Hughes, and Matthew W. Salzer. 2007. "Medieval drought in the Upper Colorado River Basin." *Geophysical Research Letters* 34: L10705.

Pearce, Michael J. 2007. "Balancing Competing Interests: The History of State and Federal Water Laws." In *Arizona Water Policy: Management Innovations in an Urbanizing, Arid Region*, edited by Bonnie G. Colby and Katharine L. Jacobs. Washington, D.C.: Resources for the Future Press.

Pulwarty, Roger S., Katharine L. Jacobs, and Randall M. Dole. 2005. "The Hardest Working River: Drought and Critical Water Problems in the Colorado River Basin." In *Drought and Water Crises: Science, Technology, and Management Issues,* edited by Don Wilhite. New York: Marcel Dekker.

REF, YOOS River

JOHN GIALANELLA

MFA, Visual Communication, School of Art, 2012

Interdisciplinary Seminar: Critical Issues in Art + Design, 2010

Walking. Searching. Thinking. Collecting.

Last fall I spent time wandering the Rillito. The once flowing, and now dry, riverbed became a sort of muse. It helped me to clear my mind and further contemplate my artistic process and my goals. A by-product of the time spent in the river is a collection of man-made objects that have been discarded, lost, or just forgotten. I think about what they are, what their purpose was, who they belonged to, where the former owners are now, and why the objects became unwanted. Sometimes these objects take on a new life and find their way into my artwork. Sometimes they just sit on my table collecting dust, once again buried and forgotten.

The time I spent in the river acted as an exercise in connection. The moonscape appearance of the riverbed stirred a feeling of abandonment, of lost contact with other humans, until trace evidence was found of their existence. The evidence of others brings comfort and a connection, while simultaneously raising more questions.

I've always been curious about what others may think about these misplaced items. Do they have the same effect on others as they do me? *REF, YOOS River* is a kit for others to interact with, to help me answer this question.

RIGHT *REF, YOOS River* | found objects | 2010

OPPOSITE PAGE *REF, YOOS River* | mixed media | 6" x 10" x 8" | 2010

REF, YOOS RIVER: JOHN GIALANELLA

Expedition

CAMDEN HARDY

MFA, Photography, School of Art, 2012

Interdisciplinary Seminar: Critical Issues in Art + Design, 2010

The Pantano Wash is a dry river that occasionally carries floodwater north through the eastern side of Tucson, Arizona, feeding into the Rillito. It serves many tangential functions, such as providing a temporary living space for transients and a place for recreation. Floods in the Pantano Wash are violent; the water often forces evidence of human presence downstream. Such a disruption of the terrain renders interpretation of this place's cultural history extremely difficult, if not impossible.

On September 17, 2010, I embarked on an expedition of the Pantano Wash. The mission was to walk all 22 miles of the dry riverbed, upstream, in an attempt to gain a better understanding of this socially charged space through first-hand experience. Photographic artifacts, which represent the end of one mile and the beginning of the next, were created using pinhole cameras and a portable processing tent. Discarded items were found along the way and cataloged into lists. Together, they offer a contemplative look at many of the unanswered questions that arose regarding the implications of human presence in the wash. The project culminated in a solo exhibition including the photographic artifacts, a video of their creation, lists of objects found, journal excerpts, the pinhole cameras and processing tent used, and the shoes worn throughout the expedition.

RIGHT *Expedition* | video stills (full video, http://vimeo.com/23516210) | 2010

Discarded Items Found

MILE 1
3 shopping carts
1 shoe
1 desk chair
1 half surfboard

MILE 2
6 plastic cups/water bottles
3 shopping carts
2 pot lids
1 red golf ball
1 spatula
1 cinder block
1 dishwasher
1 vacuum cleaner
1 empty Christmas cookie can
1 garbage bag full of teddy bears

MILE 3
4 water bottles
3 shopping carts
2 sleeping bags
1 tent, assembled
1 plastic cup
1 unidentified covered object
1 computer monitor
1 Budweiser can
1 green oil drum
1 basketball
1 mini Jack Daniels bottle

MILE 4
4 shopping carts
3 water bottles
2 beer cans
2 white ceramic plates, broken
1 shoe
1 hub cap
1 tire
1 milk crate
1 red metal lunchbox
1 empty vodka bottle
1 sleeping bag
1 camp site
1 gold leaf-shaped bowl
1 telephone wire
1 bare human footprint

MILE 5
11 shopping carts
4 water bottles
3 beer cans
3 mattresses
3 couches
2 pillows
1 tire
1 bottle grapefruit-scented soap

MILE 6
13 beer cans
6 water bottles
6 mattresses
3 shopping carts
2 shirts
2 pairs of pants
1 pair of gym shorts
1 shoe
1 "full-blown" camp site
1 Folgers coffee can
1 shotgun shell
1 rusted car
1 side view mirror
1 tire
1 Family Lawyer study software disc

MILE 7
2 "big-ass" plastic tubes
1 plastic bag full of weeds
1 empty sandbag

MILE 8
2 cinder blocks
1 light pole, formerly standing at about 60 degrees, snapped in a flood
1 inverted car, almost completely buried in dry mud
1 windshield shade
1 picnic bench
1 possible grave site

MILE 9
1 car floor mat
1 large wooden spoon
1 unidentified rusty metal object
1 washing machine top

MILE 10
4 chairs
2 nondescript metal objects
1 motor
1 sewer grate
1 pink blanket

MILE 11
1 party site, 14 beer cans spread outward from fire pit
1 cinder block
1 large metal sheet
1 blue, rusted metal object
1 orange duffel bag
1 truck bumper
1 teddy bear paw
1 shirt
1 plastic pipe
1 rusted metal bucket, unsuitable for pinhole camera
1 gallon jug of what appears to be urine
1 metal rake, no handle

MILE 12
4 unidentified rusty metal objects
3 cinder blocks
1 tire
1 tarp

MILE 13
20 cinder blocks
1 blue rope
1 towel

MILE 14 — no access

MILE 15
5 shotgun shells
2 tires
2 water bottles
1 beer can
1 license plate with puzzling slogan: "Real trucks don't need spark plugs"
1 "No Trespassing" sign, facing the opposite direction of our walk

MILE 16
1 plastic tub missing its bottom
1 beer can
1 water bottle
1 shotgun shell

MILE 17
1 child-sized shoe

MILE 18 — no items found

MILE 19 — no items found

MILE 20 — no items found

MILE 21
1 plastic pipe
1 black tarp
1 tire
1 camp site

MILE 22
1 golf ball

MILE 1

The sand was deep, and it was very hot. And I had to double back to the car. Why didn't I plan this better?

MILE 0

The surfboard on the ground to my left is an oddly appropriate way to start this journey.

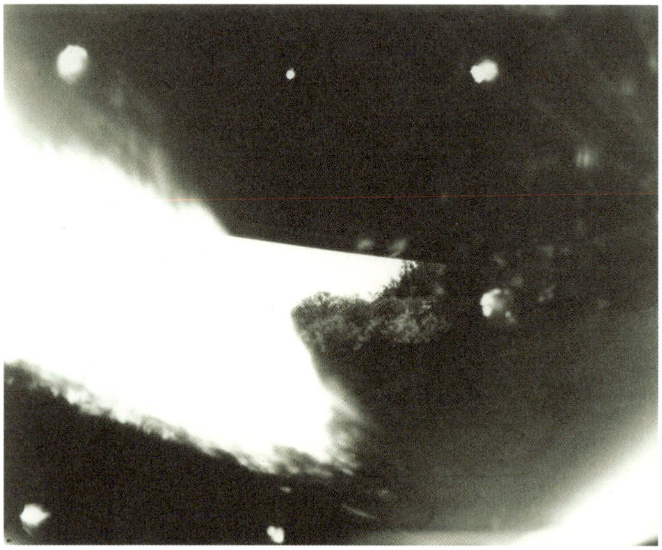

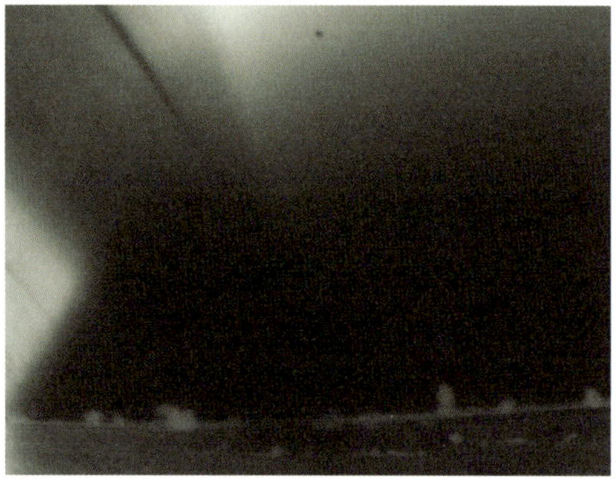

MILE 4

I suddenly found myself in a staring contest with an abnormally large bobcat looking for food. I lost.

MILE 16

A field of grass in the middle of the wash. This is one of the most surprising things I've encountered so far.

TOP LEFT *Mile 0* | silver gelatin contact print from pinhole negative | 8″ x 10″ | 2010

TOP RIGHT *Mile 1* | silver gelatin contact print from pinhole negative | 4″ x 5″ | 2010

BOTTOM LEFT *Mile 4* | silver gelatin paper negative | 8″ x 10″ | 2010

BOTTOM RIGHT *Mile 16* | photogram of sand on silver gelatin paper | 4″ x 5″ | 2010

OPPOSITE PAGE *Expedition* | Lionel Rombach Gallery | The University of Arizona | 2011

Sweetwater Journal

ROBERT LONG

MFA, Visual Communication, School of Art, 2013

Interdisciplinary Seminar: Critical Issues in Art + Design, 2010

Sweetwater Wetlands is a man-made habitat, park, and wastewater treatment facility in Tucson built in the late nineties. Secondary effluent is pumped through Sweetwater's ponds, where plants and microbes remove chemical contaminants. The plants put dissolved oxygen into the water, which is necessary for the base of the aquatic food chain. After being filtered through the wetlands, the water flows into recharge basins, where it is purified by ultraviolet light before it soaks into the ground to replenish the aquifer. This water can then be pumped from the ground for use in Tucson's reclaimed water system.

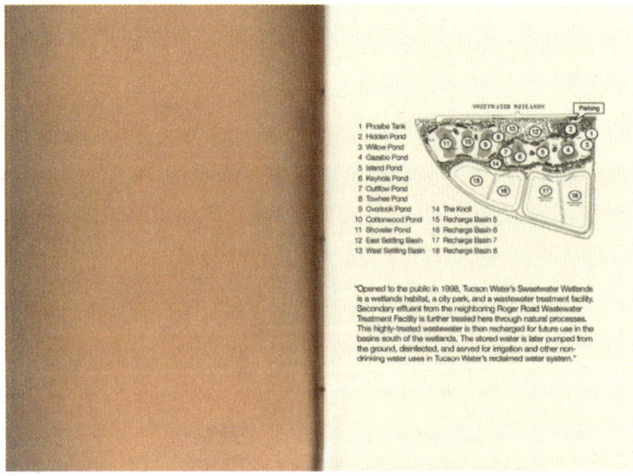

I kept a visual and textual journal of my visits to this habitat in a 6-by-4-inch handbound book. Using ballpoint pen and graphite, I documented my experiences through annotated illustrations. I surveyed the landscape and its small inhabitants, and visually dissected spatial volumes occurring in the wetland: the enclosed micro-lagoons created by both cattails and a low canopy overhead; the vascular canals within a tree bough; the hollowed-out thorax of a downed moth, which may have been shelter for an even smaller creature. The need to investigate draws me close to miniscule subjects and allows me to empathize with them.

A narration both technical and humorous threads its way through my sequential account, describing the conditions under which the drawings were made. While describing nature, I take note of my presence in relation to my subjects and record details of my circumstances (e.g., "a fly just went up my nose"). Participation of the senses and inclusion of tangential thoughts offer a more intimate view than that of a traditional scientific illustrator, who may fend off rather than embrace distraction.

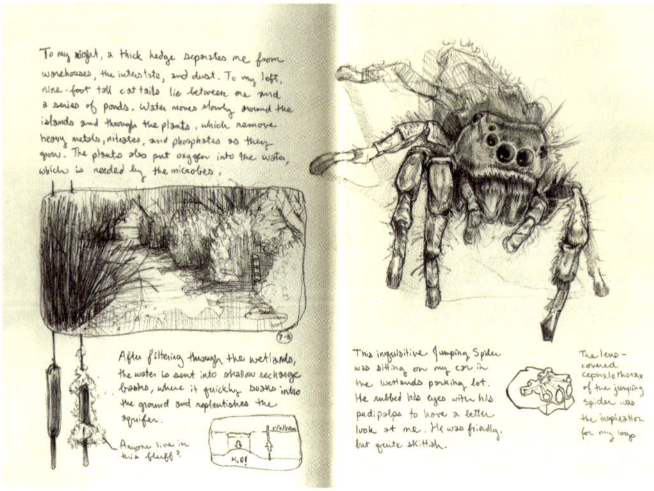

ALL IMAGES Pages from *Sweetwater Journal* | Ballpoint pen and graphite | 6″ x 4″ | 2010

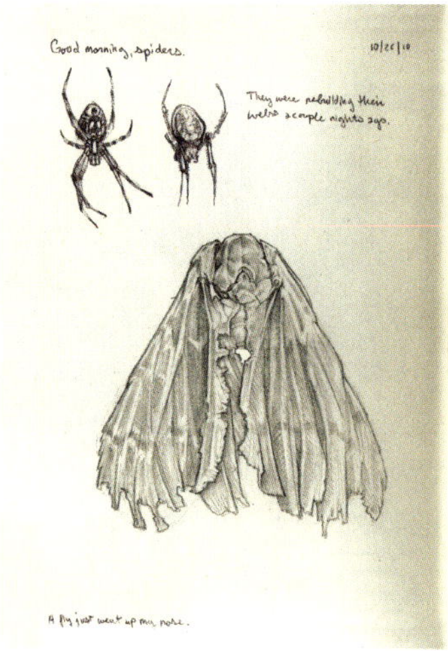
Good morning, spiders.
10/26/10
They were rebuilding their webs a couple nights ago.
A fly just went up my nose.

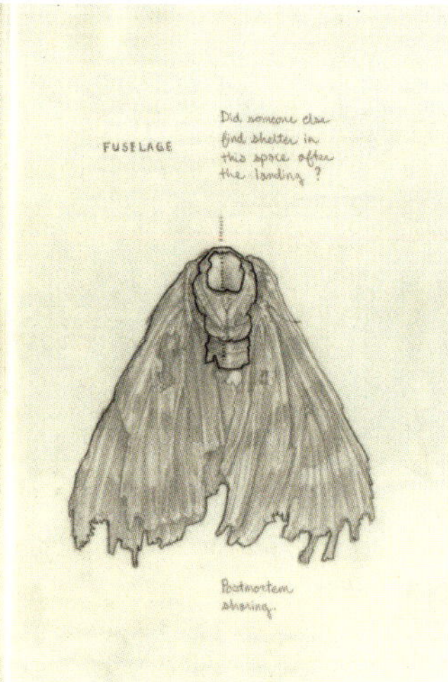
FUSELAGE
Did someone else find shelter in this space after the landing?
Postmortem shoring.

11/2/10
No spiders at the entrance today, but there was a little grey bug on the bridge. He had piercing mouthparts and liked playing dead.
Dried, collapsed reeds mark where the water level had been. In the dead leaves, an expanse of funnel-shaped spider webs catch the light and become visible.
③
Willow
To keep the sun at my back, go clockwise around the ponds in the morning and counterclockwise in the afternoon.

Fallen Willow between ③ and ④ with termite tunnels, but its crawling with little black ants. Several branches of the tree seem to still be alive and well.
11/9/10
④
Upon closer inspection, I don't think they're termite tunnels at all, but those of carpenter ants.

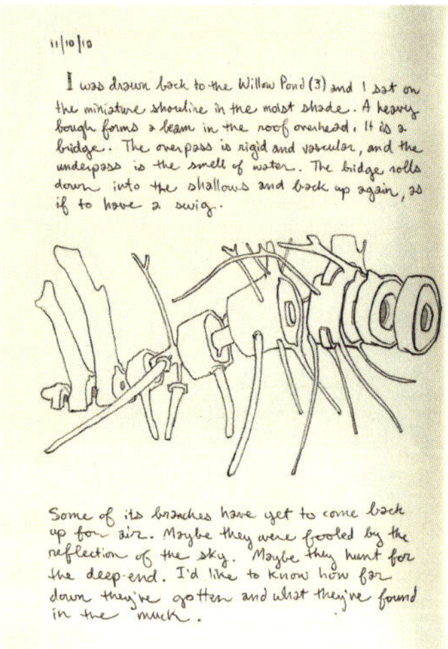
11/10/10
I was drawn back to the Willow Pond (3) and I sat on the miniature shoreline in the moist shade. A heavy bough forms a beam in the roof overhead. It is a bridge. The overpass is rigid and vascular, and the underpass is the smell of water. The bridge rolls down into the shallows and back up again, as if to have a swig.
Some of its branches have yet to come back up for air. Maybe they were fooled by the reflection of the sky. Maybe they hunt for the deep-end. I'd like to know how far down they've gotten and what they've found in the muck.

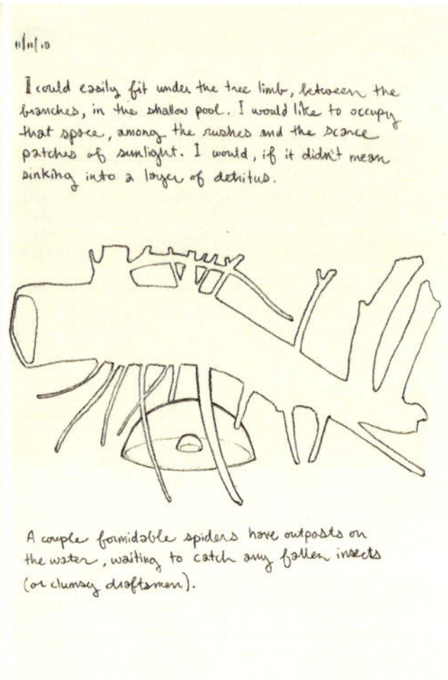
11/11/10
I could easily fit under the tree limb, between the branches, in the shallow pool. I would like to occupy that space, among the rushes and the scarce patches of sunlight. I would, if it didn't mean sinking into a layer of detritus.
A couple formidable spiders have outposts on the water, waiting to catch any fallen insects (or clumsy draftsmen).

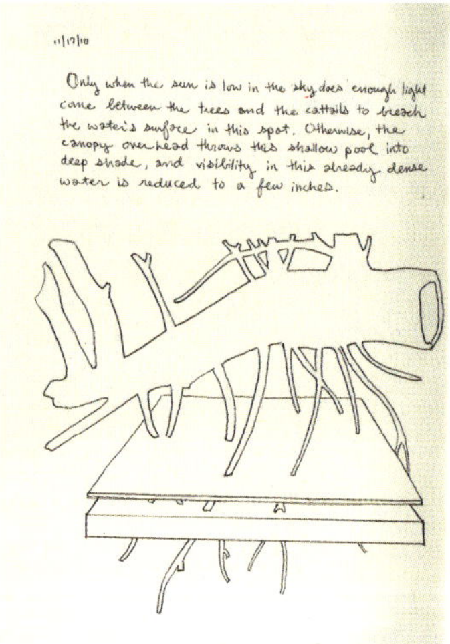
11/17/10
Only when the sun is low in the sky does enough light come between the trees and the cattails to breach the water's surface in this spot. Otherwise, the canopy overhead throws this shallow pool into deep shade, and visibility in this already dense water is reduced to a few inches.

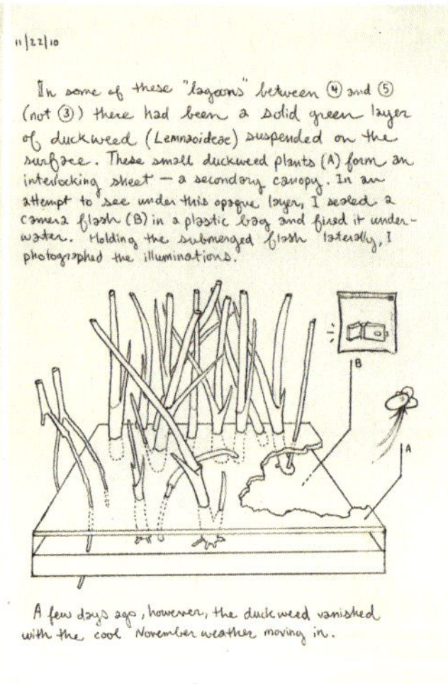
11/22/10
In some of these "lagoons" between ④ and ⑤ (not ③) there had been a solid green layer of duckweed (Lemnaoideae) suspended on the surface. These small duckweed plants (A) form an interlocking sheet — a secondary canopy. In an attempt to see under this opaque layer, I sealed a camera flash (B) in a plastic bag and fixed it underwater. Holding the submerged flash laterally, I photographed the illuminations.
B
A
A few days ago, however, the duckweed vanished with the cool November weather moving in.

A free-floating cattail rhizome sails slowly
around me. There's another cluster afloat behind
it, and two Coots are pecking at it, probably
eating the algae off of the roots. Looking
North across the pond, I can see the other
side of the ant-infested tree from 11/9/10.
But only while standing here, held out over
the water can I see my breath.
11/29/10
4
N
5:08 pm

Imagination and Invention: Romanticism and the Life of a Dry River

ALLISON DUSHANE

"Do you think literature can actually save the world?" As a scholar of British Romantic-era literature (roughly 1790–1830) who teaches courses that engage with the interactions between literature, scientific practices, and environmental concerns, I often hear this question. And it's a legitimate one. In a time of increasing environmental and economic crisis, why should we be spending increasingly scarce resources, time, and money reading, discussing, and writing about literature written 200 years ago? What could a group of poets, essayists, and novelists writing without any threat of water scarcity have to teach us about the challenges that face us in a time of tremendous environmental change, disappearing habitats, decreasing and uncertain future prospects? How might they respond, for example, to the phenomenon that inspired the Rillito River Project installation, *The Water Table*, which provided a striking visual representation of the rapidly decreasing water level of the Rillito River that has accompanied Tucson's increasing population, transforming what was once a vibrant riparian habitat to a desolate, dry riverbed? I hope to shed some light on these questions through a brief consideration of some of the key texts that have come to represent Romanticism as a literary-philosophical movement along with some lesser-known ones, focusing on an exploration of how thinkers in this time period define and deploy the linked faculties of imagination and invention to establish—and potentially transform—relationships between human beings, material things, and the world we call "Natural."

In 1789, William Wordsworth began composing the verses that comprise the text known as the "Two-Part Prelude." In this early version of the fourteen-book *Prelude*, a poem that would go on to be celebrated as the great autobiographical epic of the Romantic Age, Wordsworth, in traditional fashion, turns to a muse. Instead of calling upon the Classical muses in the tradition of Homer, Virgil, and Dante, or upon the Christian spirit of prophecy invoked by his beloved Milton in *Paradise Lost,* Wordsworth turns to a more concrete and material figure: a river. Specifically, he calls upon the river Derwent, which winds its way through the mountains of his birthplace in Cumberland, England, and serves as both a literal source of and metaphor for the imagination as it was conceived by Romantic-era thinkers:

> For this didst thou
> O Derwent, travelling over the green plains
> Near my "sweet birth-place," didst thou beauteous Stream
> Make ceaseless music through the night and day,
> Which with its steady cadence tempering
> Our human waywardness, composed my thoughts
> To more than infant softness, giving me,
> Among the fretful dwellings of mankind,
> A knowledge, a dim earnest of the calm
> Which Nature breathes among the fields and groves?
> (First Part, 6–15)[1]

In Wordsworth's account, the river Derwent plays a crucial role in the formation of the mind of this poet who would go on to proclaim himself a Prophet of Nature—one perhaps guided by destiny or intention, as he hints in the rhetorical question "For this." It may seem the height of vanity to devote a project that would grow to fourteen books of poetry written in an epic tradition to the genesis and development of one's own self, or perhaps not, given the individualist impulse of our own culture.

Yet, there is another way to read *The Prelude*, in which the river Derwent is one of many indications of the presence and activity of the nonhuman entities that mysteriously shape and connect individual human lives, such as Wordsworth's formative childhood exploits in his rural home, and landmark historical events such as the French Revolution. An interpretation of Wordsworth's claims to the status of Prophet depends upon a more nuanced understanding of what Romantic-era writers mean when they talk about the relationship between poetry and prophecy, and, more crucially, what exactly they mean—and don't mean—by Nature. The latter question, and the ways in which the answers to that question have a significant bearing on our present environmental challenges, has given rise to a rapidly expanding and interdisciplinary field of literary study: ecocriticism. As it grew out of an environmentalist impulse, ecocriticism began with a tendency to read literary texts as a mode of recovering or engendering an appreciation of the natural world that would instrumentally lead to its preservation. The unfortunate and often unrecognized side effect of such an approach, however, is the way in which such readings set up Nature as something graspable that nevertheless always exceeds our grasp, something to retreat into as a refuge from the stresses and destructive qualities of modern or city life, something that actually *exists* in a stable and reliable way, fleetingly accessible to but ultimately outside the reach of human thought. As the author of the *Cambridge Introduction to Literature and the Environment* puts it:

> The initial impetus of modern ecocriticism...lies some two centuries in the past, in a broadly romantic tradition of opposition to the destructive tendencies of enlightenment ideals of the conquest of nature, the market based economy and industrialism...romantic with a small r, to name continuing and deeply engrained modes of thought that oppose industrial society with ideas of 'nature' and 'the natural' as modes of secular redemption.[2]

In many ways, Romantic-era writing is marked by this kind of thinking, which is often what is meant when one calls a thinker "romantic": naïve, idealistic, skeptical (or outright ignorant) of the material concerns of daily life. Yet, at the same time that Romantic-era writing contributes to the construction of a "romantic with a small r" conception of nature, it also provides a rich source of materials from which to critique, rethink, and revise such conceptions. Many of these texts have come more sharply into focus—and been given wider availability and attention in the classroom—as a result of these interdisciplinary efforts.

> The faculty that allows human beings to "invent," literally and metaphorically, is embedded in our material entanglement with the rest of the natural world.

Take, for example, the poetry and prose of Erasmus Darwin, best known today as grandfather to Charles. A physician by profession, Darwin possessed an incredibly polymathic mind and left a lasting mark on disciplines such as medicine, botany, chemistry, and geology. He was also the most popular poet in England in the 1790s. His first major literary publication in 1789, *The Loves of the Plants*, was written to illustrate the Linnaean system of classification, using heroic couplets, the conventions of epic, and references to history and mythology to animate the adventures of a host of personified plants. *The Loves of the Plants* became Part II of his longer project, *The Botanic Garden,* when an edition bound with Part I, *The Economy of Vegetation*, came out three years later. Part I draws upon a wider base of Darwin's knowledge of science and technology, shifting from an illustration of classification to an explanation of "the physiology of Plants" and "the operation of the Elements." What makes *The Botanic Garden* such a singular text isn't necessarily the fact that the verse focuses on scientific subject matter, but that they are accompanied by over 100,000 words of prose, "Philosophical Notes" that link the images and actions in the poem to the work and writing of contemporary scientists, inventors, and artists through extensive citation and commentary. For example, consider this passage depicting the origins of the universe through a proto-big bang theory, for which Darwin came under attack for its early vision of an evolutionary and self-sustaining nature:

> "LET THERE BE LIGHT!" proclaim'd the ALMIGHTY LORD,
> Astonish'd Chaos heard the potent word;—
> Through all his realms the kindling Ether runs,
> And the mass starts into a million suns;
> Earths round each sun with quick explosions burst,
> And second planets issue from the first;
> Bend, as they journey with projectile force,
> In bright ellipses their reluctant course;
> Orbs wheel in orbs, round centres centres roll,
> And form, self-balanced, one revolving Whole.[3]

Darwin's footnote to this passage explains its origin in the work of astronomer (and noted composer) William Herschel, and directs readers to Herschel's work on star systems in the *Philosophical Transactions of the Royal Society*, Britain's premier publication for scientific papers. As for the driving force behind the origin of the universe and the process of its development, Darwin muses: "We can have no idea of a natural power, which could project a Sun out of Chaos...of the power of which under immeasurable degrees of heat, and compression, we are yet ignorant" (1.1.105n). Despite that basic ignorance of nature's power, Darwin proceeds to narrate nature's processes in tandem with an account of man's efforts to understand and harness them for human ends. These include a long note on the invention of the steam engine and its possible applications, ranging from the immediately practical: "raising water from mines, blowing bellows to fuse the ore, supplying towns with water, grinding corn," to the futuristically fanciful: "there seems no probable method of flying conveniently but by the power of steam, or some other explosive material; which another half century may probable discover" (1.1.254n).

Darwin proclaims in the advertisement to *The Botanic Garden* that his aim is to "enlist Imagination under the banner of Science," and in its efforts to do so it might be considered a work that bridges the prevailing methods and sensibilities of the Age of Enlightenment and the Romantic era. As an Enlightenment text, it praises the scientific method and hails the instruments that enable the construction of scientific facts, professing an exuberant faith in the potential that experimental technologies and the products of industry offered for the progress of human society. As a Romantic one, it celebrates art as a distinct way of knowing the world. The poem itself, then, can be read as an instrument designed to instruct—and delight—members of a reading public, writing them into the discourse of the Enlightenment and recruiting the powers of the imagination to its cause. *The Botanic Garden* promises to include its readers in the wider scientific community, spreading the spirit of invention and experiment beyond the walls of official institutions like the Royal Society. The text is, from a contemporary perspective, interdisciplinary in terms of its subject matter, using poetry as a forum to discuss not only scientific topics appropriate to the discussion in polite society such as the classification of plants, but to open up more controversial debates about the origin of the universe and the relatively recent entry of human life into its history. What makes it particularly singular from our point of view, however, is the fact that it not only uses poetry to deal with scientific material, but that it also takes advantage of certain scientific conventions in the presentation of its poetry, specifically the citation and comparison with previous experimental findings in regulating its "truth." That is, *The Botanic Garden* is relevant to contemporary discussions of interdisciplinarity in the academy because it reminds us that breaking boundaries is not necessarily about bringing scholars with distinct approaches together to discuss a specific topic, but about allowing them to interact in such a way that their work is affected at the level of disciplinary training, of method.

For Darwin, as well as many other Romantic-era writers, the line between viewing the world from an Enlightenment and a Romantic perspective, between celebrating invention and praising the imagination, was not drawn as clearly as the lines between the sciences and the humanities today. In his two-part medical treatise, *Zoonomia*, Darwin opens his then-revolutionary treatment of diseases as systemic disruptions borne of the relationships between bodily systems with a consideration of the relationship between matter and mind. In his account, invention is a faculty of the mind inextricably linked to materiality, which he analogizes to a machine:

> Invention is an operation of the sensorium, by which we voluntarily continue to excite one train of ideas, suppose the design of raising water by a machine; and at the same time attend to all other ideas, which are connected with this by every kind of catenation; and combine or separate them voluntarily for the purpose of obtaining some end.[4]

In other words, the faculty that allows human beings to "invent," literally and metaphorically, is embedded in our material entanglement with the rest of the natural world, and identifies the point at which an accumulation of basic sensory perceptions, assembled into complex ideas, can be harnessed into specific goals and purposes. In a section detailing his theory of human reproduction and embryological development, "Of Generation," he goes on further to claim that human identity itself is not static or given, but shaped through an ongoing material process that he links with the evolutionary and self-generating vision of nature he proposed in *The Botanic Garden:*

> ...would it be too bold to imagine, that in the great length of time, since the earth began to exist, perhaps millions of ages before the commencement of the history of mankind...that all warm-blooded animals have arisen from one living filament, which The Great First Cause endued with animality, with the power of acquiring new parts, attended with new propensities, directed by irritations, sensations, volitions, and associations; and thus possessing the faculty of continuing to improve by its own inherent activity, and of delivering down those improvements by generation to its posterity, world without end! (I.XXXIX.8)

According to Darwin, human beings are constant works in progress, subject to nature's mysterious power while also being responsible for harnessing that power towards its continual improvement.

Darwin's poetry and ideas were incredibly influential on the Romantic-era writers that receive the most classroom attention today, many of whom met or knew him personally in addition to possessing and remarking upon his work, including Wordsworth, William Blake, John Keats, Percy Shelley, Mary Shelley, and the foremost theorist of the Romantic imagination, Samuel Taylor Coleridge. They would turn a critical eye towards Darwin's optimistic faith in industry and linear progress in the manner that gave rise to contemporary ecocritical perspectives. However, their conceptions of a constantly changing, developing, dynamic nature all are informed heavily by their knowledge of science, and even personal relationships with particular scientists.[5] Romantic-era conceptions of the imagination often link the formation of the self with the operations of the material world. For example, Coleridge's famous account of the imagination in his philosophical autobiography, *Biographia Literaria,* draws from a similar line of thinking as Darwin's account of the faculty of invention. In his formulation, the imagination has "primary" and "secondary" aspects: "The primary IMAGINATION I hold to be the living Power and prime Agent of all human Perception, and as a repetition in the finite mind of the eternal act of creation in the infinite I AM." As in Darwin's account of the origin of the universe, we see the reference to a mysterious "living Power," this time one that runs through all of creation and lies at the source of human faculties. Coleridge then proposes the secondary imagination as:

> ...an echo of the former [primary imagination], co-existing with the conscious will, yet still as identical with the primary in the kind of its agency, and differing only in degree... It dissolves, diffuses, dissipates, in order to re-create; or where this process is rendered impossible, yet still at all events it struggles to idealize and to unify (337).

The imagination, then, is a faculty that links thinking beings to the material world through a "living Power" that exerts itself over and through human life, allowing human beings to participate in the inextricably linked activities of creating art and inventing things. Coleridge calls the imagination the "Esemplastic Power," coining the word esemplastic from Greek and German roots to convey the sense of "unifying" or "forming into one." In one of his most influential poems, "The Eolian Harp," he contemplates the musical instrument named in the title, a lyre that creates sound through exposure to passing wind, and poses a question similar to the one Darwin posed in *Zoonomia*:

> And what if all of animated nature
> Be but organic Harps diversely fram'd,
> That tremble into thought, as o'er them sweeps
> Plastic and vast, one intellectual breeze,
> At once the Soul of each, and God of all? (45–49)

What if the intellectual activity of human beings is inseparable from the activity of "animated nature"? What if nature has more power over us than we have over it? What if we are unable to fully quantify or tame something so "plastic and vast"? Romantic-era writing is littered with essays, poetry, prose, and fiction that are dedicated to posing such questions.

> The line between viewing the world from an Enlightenment and a Romantic perspective, between celebrating invention and praising the imagination, was not drawn as clearly as the lines between the sciences and the humanities today.

The increasingly blurry line between human and nonhuman agencies—between the activity of natural entities such as rivers and breezes, man-made objects such as steam-engines and harps, and the creative and intellectual activities of the mind—is what prompts Wordsworth to cast poetry as prophecy. The advertisement to one of the volumes of poetry that has come to characterize the Romantic era, the 1798 *Lyrical Ballads*, copublished anonymously by Wordsworth and Coleridge, tells the reader that the "majority of the following poems are to be considered as experiments."[6] Departing from the elevated style and Classical forms that dominated much of the poetry of the eighteenth century, *Lyrical Ballads* featured poems written in "the language of conversation in the middle and lower classes of society," including the perspectives of a "Female Vagrant," a "Mad Mother," an "Idiot Boy," a "Forsaken Indian Woman." Such figures are characterized by their displacement from or refusal to conform to the standards of living and behavior of the educated class that comprised Wordsworth and Coleridge's audience, challenging the "natural," or "romantic with a small r" ideal of nature that readers expected to find enshrined in poetry. As one reviewer for the British periodical *Critical Review* put it, "the 'experiment,' we think, has failed, not because the language of conversation is little adapted to 'the purposes of poetic pleasure,' but because it has been tried upon uninteresting subjects."[7] With careful attention to the changes wrought on a rural community and its inhabitants by political and economic forces, Wordsworth and Coleridge refuse to ignore the "Vagrant," the "Mad," the "Idiot" and the "Forsaken," writing these marginalized characters into the fabric of the changing landscape. *Lyrical Ballads* creates a sense of place that draws from a blend of human and nonhuman landmarks in poems like "The Waterfall and the Eglantine," "The Oak and the Broom," "The Thorn," and "Rural Architecture." Later, Wordsworth would go on to attach his name to two more editions of *Lyrical Ballads*, in 1800 and 1802, for which he wrote an expanded Preface more fully explaining the aims of his poetic project. He writes:

> If the labours of men of Science should ever create any material revolution, direct or indirect, in our condition, and in the impressions which we habitually receive, the Poet will sleep then no more than at present, but he will be ready to follow the steps of the man of Science, not only in those general indirect effects, but he will be at his side, carrying sensation into the midst of the objects of the Science itself. (423)

In Wordsworth's vision of the future, "[t]he remotest discoveries of the Chemist, the Botanist, or Mineralogist, will be as proper objects of the Poet's art as any upon which it can be employed" (423). Nature is something far more complex and dynamic than the idealized space it can seem to be without looking at the bigger picture, which includes the constant activity of physical nature, nonhuman biological organisms and their encounters with and incorporation into the objects and tools that human beings invent and produce to sustain their lives. Nature emerges as something that incorporates the human and the nonhuman through their shared materiality:

> If the time should ever come when what is now called Science, thus familiarized to men, shall be ready to put on, as it were, a form of flesh and blood, the Poet will lend his divine spirit to aid the transfiguration, and will welcome the Being thus produced, as a dear and genuine inmate of the household of man. (424)

To be a "Prophet of Nature," then, is to provide a conduit between science and art, invention and imagination.

Nature emerges as something that incorporates the human and the nonhuman through their shared materiality.

To return to the original question, "Do you think literature can actually save the world?" I would answer no. That is, the problem that faces us today is not one to be "solved" by "saving" something. Rather, it is necessary to marshal all of the resources at our disposal, scientific and humanistic, towards rethinking, reimagining, and reinventing a world that recognizes the inseparability of human and nonhuman concerns. I would argue that Wordsworth's conjectures in the Preface to the *Lyrical Ballads* pose (one could say prophesy) a crucial issue facing universities today: how do we balance the specialization and resources required to produce scientific knowledge with the need to translate it into the messier realm of culture? From the perspective of literary studies, ecocriticism articulates it as a challenge: "to keep one eye on the ways in which 'nature' is always in some ways culturally constructed, and the other on the fact that nature really exists, both as the object and, albeit distantly, as the origin of our discourse."[8] In *The Ecological Thought*, Timothy Morton advocates taking apart our received ideas of "Nature" in favor of thinking through the "mesh," which incorporates the human and the nonhuman, the biological and the mechanical, the ugly and the beautiful into a way of thinking that embraces randomness and contingency instead of retreating into an idealized past to construct a utopian future. He argues that ecology "isn't just about global warming, recycling, and solar power—and also not just to do with everyday relationships between humans and nonhumans. It has to do with love, loss, despair, and compassion. It has to do with depression and psychosis."[9] It has to do, in other words, with what literary forms and other means of artistic expression make materially present, with their potential to grant "flesh and blood" to scientific ideas, practices, and nonhuman entities. Just as Wordsworth, over 200 years ago, called on the river Derwent as inspiration for recording the life of an individual, a community, and an era, so the collection of visual artists, writers, architects, scientists, and activists that have come together through *Ground/Water* call upon the dry bed of the Rillito to rethink, re-imagine, and reinvent the life of a desert landscape.

NOTES

1 All references to Romantic-era poetry and prose, unless otherwise indicated, refer by line or page number to *The Age of Romanticism: Broadview Anthology of British Literature*, Vol. 4. Eds. Joseph Black, et al. Toronto: Broadview, 2006.

2 Timothy Clark, *The Cambridge Introduction to Literature and the Environment*, (Cambridge: Cambridge University Press, 2011), 14.

3 Erasmus Darwin, *The Botanic Garden: A Poem in Two Parts* (London: J. Johnson, 1791), 1.1.103–12.

4 Erasmus Darwin, *Zoonomia; or, the Laws of Organic Life* (Boston: Thomas and Andrews, 1803), I.XV.3.

5 The scholarly literature on the relationship of these authors to scientists is incredibly vast, but an excellent introduction written for a more popular audience would be Richard Holmes, *The Age of Wonder: How the Romantic Generation Discovered the Beauty and Terror of Science* (New York: Pantheon, 2009).

6 William Wordsworth and Samuel Taylor Coleridge, *Lyrical Ballads 1798 and 1800*, ed. Michael Gamer and Dahlia Porter (Toronto: Broadview, 2008), 47.

7 Ibid. pp. 149–150.

8 Greg Garrard, *Ecocriticism* (New York: Routledge, 2004), 10.

9 Timothy Morton, *The Ecological Thought* (Cambridge: Harvard University Press, 2010), 2.

REFERENCES

Black, Joseph, Leonard Conolly, Kate Flint, Isobel Grundy, Don LePan, Roy Liuzza, Jerome J. McGann, Anne Lake Prescott, Barry V. Qualls, and Claire Waters., eds. 2006. *The Age of Romanticism: The Broadview Anthology of British Literature,* vol. 4. Toronto: Broadview Press.

Clark, Timothy. 2011. *The Cambridge Introduction to Literature and the Environment*. Cambridge: Cambridge University Press.

Darwin, Erasmus. 1791. *The Botanic Garden: A Poem in Two Parts*. London: J. Johnson.

Darwin, Erasmus. 1803. *Zoonomia; or, the Laws of Organic Life.* Boston: Thomas and Andrews.

Garrard, Greg. 2004. *Ecocriticism*. New York: Routledge.

Holmes, Richard. 2009. *The Age of Wonder: How the Romantic Generation Discovered the Beauty and Terror of Science*. New York: Pantheon.

Morton, Timothy. 2010. *The Ecological Thought.* Cambridge: Harvard University Press.

Wordsworth, William, and Samuel Taylor Coleridge. 2008. *Lyrical Ballads 1798 and 1800.* Edited by Michael Gamer and Dahlia Porter. Toronto: Broadview Press.

Sweetwater Wetlands, 2010

DANIEL CHEEK

MFA, Photography, School of Art, 2013

Interdisciplinary Seminar: Critical Issues in Art + Design, 2010

Sweetwater Wetlands is a project of Tucson Water to reclaim wastewater for use on golf courses, public parks, schools, and other large turf areas. The location now fulfills several different functions in addition to its original prosaic one. It has become a habitat for many birds, mammals, reptiles, amphibians, insects, and plant life. It also serves as a recreational park where people watch the wildlife and stroll around a lush green wetland, a rare opportunity in the desert. I was interested in photographing the duality of Sweetwater's existence—to show the lushness and wildness of the wetland plant life, how people experience this landscape, and the evidence of its mechanical purpose.

ALL PHOTOGRAPHS *Sweetwater Wetlands, 2010* | silver gelatin prints | 8″ x 10″ | 2010

SWEETWATER WETLANDS, 2010: DANIEL CHEEK

SWEETWATER WETLANDS, 2010: DANIEL CHEEK

Talking through Objects: Multidisciplinary Dialogues with "Things"

PAUL ROBBINS

Biking past the Rillito River on a Saturday morning in September, shortly after artists and scholars had helped to convene a highly successful Bat Night, I was reminded of the many kinds of people who had converged in this interaction: essayists and political ecologists, students studying architectural design and photography, experts in water and those who have only begun to think about it. More tellingly, I was struck by the many paths that this diverse community had followed to reach this combined work, and the simplicity of the technique that had united them. This multidisciplinary dialogue, as productive as it is improbable, had been assembled simply by having all the participants follow a dry riverbed to one another. What does this tell us about fostering interdisciplinary dialogue?

This essay briefly explores the way interdisciplinary activities are supposed to happen, through the creation of a unifying set of problems, questions, and models, amongst collaborating experts, compared to how it often actually does happen: through negotiating misunderstandings amongst jumbled clusters of people, over the nature and condition of real-world objects and things, like the Rillito. This difference, between an idealized protocol and an experienced "muddling through," is one that bears reflection and raises questions about how to begin and sustain communities of conversation. Accepting that under many circumstances the central foci of interdisciplinary conversation are material objects and landscapes (elk, wells, canyons, mosquitoes, fences, rocks...bats) means that translation across sometimes vast chasms of mutual incommunicability may sometimes be indirect, and ultimately mediated by other actors. This may therefore direct us to both a more modest but more immersive way to imagine communities of research assembled around objects rather than problems, grand challenges, or bundles of concepts. Accepting this further invites such objects of concern to announce themselves through their relationship to diverse communities, in their specificities, and on their own terms, with all of the thorny ethical, normative, and political implications this might hold.

Convening eclectic communities around a dry riverbed, therefore, is a protocol that is guaranteed to produce revelation, and possibly new forms of politics, even if it provides no assurance of eventual perfect mutual comprehensibility. As such, it resembles more closely interdisciplinary practice "as it is," rather than how we may wish it to be, and holds out a promising alternative to grand interdisciplinary synthesis. It also shows the way art and the arts, in particular, can help reveal the very grounded and material strands that bind specialists to one another. In this way, artists become catalytic in creating conversations through objects of concern.

INTERDISCIPLINARY RESEARCH AS IT IS IMAGINED: THE SOCIABILITY OF PEOPLE

Trying to foster interdisciplinary research is a task that countless agencies and institutions have taken as central to their mission.

Whole universities have been reordered in recent years away from disciplinary units into integrative centers and schools (and so the proliferation of interesting new nondisciplines, like sustainability science, cognitive science, human-computer interaction, and so on). These efforts are admirably rooted in the critical revelation that the most exciting innovations come at the edges, rather than the cores, of traditions and habits of thought.

More specifically however, two central assumptions are mobilized in efforts to create and encourage such conversations. First, the conceived channel for interdisciplinary efforts is one of integrating technologies of knowledge: ideas, methods, and concepts. Second, the mode of organization is one that focuses on fostering human sociability: interaction between researchers. In the first case, as understood by the National Academy of Sciences in the United States, interdisciplinary research is defined as:

> a mode of research by teams or individuals that integrates information, data, techniques, tools, perspectives, concepts, and/or theories from two or more disciplines or bodies of specialized knowledge to advance fundamental understanding or to solve problems whose solutions are beyond the scope of a single discipline or area of research practice. (Committee on Facilitating Interdisciplinary Research 2004)

One can imagine what is intended here by way of example. Consider the challenge of biodiversity conservation in the landscapes of the Amazon, where farming is ongoing. Traditional specialists in agronomy and those in conservation biology might be convened to "solve the problem" of reconciling production with plant and animal conservation. They might be expected to meet and share concepts and methods with one another until such a time as they achieve a "breakthrough," and some integrated concept or method appears.

Yet in such a way of working, what is actually integrated? According to this definition, it is "information, data, techniques, tools, perspectives, concepts, and/or theories." The technologies of specialists are supposed to mix, and some emergent property is imagined to arise through integration. Though such an approach may succeed, its limits should also be evident. How would we know in advance what categories or specialists ought to be in dialogue? What might guide them, besides trial and error, to a convergent mode of analysis? As a matter of scientific habit, the tendency in such cases is to design some kind of single or master model, into which the single categories of the researchers are entered or integrated. In this case, agronomists are expected to provide rates of inputs and yields of output, for example, while conservations may measure the habitat loss from differing degrees of agricultural intensity. They reach an understanding, expressed as a predictive or interactive model, which might be considered "interdisciplinary."

Viewed this way, the prerequisite for integration is bringing the right specialists into sustained contact, under the right conditions and with the right incentives. The National Research Council (NRC) provides some further guidelines to what these conditions might be based on the reported conditions under which sustained interdisciplinary interaction is thought to occur according to leaders in this field: scholars need a chance to encounter one another, they need to be rewarded for doing so, they need money, and so on. There is nothing objectionable to these conclusions and they are worth consideration.

Yet almost no mention is made in the guidelines of the material or substantive bridge that might compel emergent discovery, the thing the researchers actually share, except a brief mention that they must have a "problem to solve." The NRC stresses the key assumption of this approach, therefore. In co-location, funding, organization, and team-building, the most important relationships in interdisciplinary research are posited to be between researchers and other researchers. Interdisciplinary work is imagined as a narrowly human social problem, therefore, one of getting people into contact and communication with other people. Critically, no acknowledgment is made here of the role of the relationships between the researchers and things other than people. This assumption prevails despite the clear and intimate importance of nonhuman objects and beings within the practice of research: elephants, bench equipment, gardens, or microbes.

Returning to our Amazonian case, it should not be surprising, therefore, that interdisciplinary work in this field has proliferated complex models, uniting diverse elements into abstracted computing environments. Data is integrated through models by socializing researchers. These models take a fragment of research product or concept from each specialist, enumerate them, and allow them to mingle in complex algorithms. Since the social world of much Amazonian interdisciplinary research is populated by agronomists and conservation biologists, it is largely unsurprising what the models come to look like. They necessarily reflect the cross-parentage of the specialists involved. Most land-cover change models come to stress the importance, notably, of the relationship between how intensive agriculture is (its agronomic "gene"), and how much habitat remains for biodiversity (its conservation "gene") (Morton et al. 2006; Walsh et al. 2008). A little of each of the specialist emerges from the mix. Mission accomplished.

INTERDISCIPLINARY RESEARCH "AS IT IS": THE SOCIABILITY OF MORE-THAN-PEOPLE

But other ways of coming to dialogue exist. Consider recent work surrounding coffee in the Amazon, an agro-ecosystem that produces a cacophony of unlikely research dialogues on biodiversity and society, which reflects a model of collaborative learning rather differently. Consider how these plantations create convergence in the work of ecologists and political economists.

For ecologists, who are interested in tracking ants and spiders through trees and plants, coffee holds an attractive allure. As a quasi-artificial ecosystem, the environment allows researchers to ask how new actors function in an ecosystem, how ecological component species can be "traded out" from more "natural" surrounding forests, and whether novel ecologies can be formed out of the mix of coffee plants and native forests. These researchers trudge through the landscape, observe interactions between different creatures and things, and observe how farmers work to weave these together (Vandermeer et al. 2010).

> A revelation is made possible because the sociability of scientists (or artists, or writers, or actors...) is not only with one another, but with the objects of the world.

At the same time, political economists are drawn to these same landscapes because they represent economic innovations in the landscape, compromises of labor time and inputs, all within the global linkages of a turbulent commodity economy. These researchers walk through these same forests, dialoguing with famers, counting resources, and tracking the ongoing experiments of local producers. These are joined by historians trying to understand struggles over land and proliferation of poverty and the leakage of value through international commodity chains (Wright and Wolford 2003).

Coffee plantations convene a population of seekers and scholars, who can't help but bump into one another, as well as local people, creatures, and plants, creating a social world around Amazonian forests that differs from the compromises of model-driven interdisciplinary work. The jagged accounts that emerge in such dialogues by no means produce a single question, model, or result, but instead create a productive juxtaposition of connections, branching outward from trees and farmers' fields, filled with ecological immediacy and political urgency. The results of this kind of work point to the terrible work coffee can do on local livelihoods, linked as the commodity is to a roller coaster of global trade, with the potential of economic disaster always looming. At the same time, the complex web of relations sewn together in these human-made forests allow producers to carve out a living in a global market, even while they garden the earth in a way that might be sustained over whole regions and across generations (Perfecto et al. 2009).

It is coffee that makes scientific worlds collide and announces itself, in the process, as a harbinger of poverty but also a route to democratic sustainability. Coffee is confronted by diverse researchers, and with it a dawning recognition of the independent power of the coffee plant itself, its ecology, and the concatenation of its surrounding communities (of farmers and insects). This revelation is made possible because the sociability of scientists (or artists, or writers, or actors...) is not only with one another, but with the objects of the world. The emergence of an interdisciplinary conversation is one that occurs through fits and starts, cross-referenced across translated languages, with communities stumbling into one another in fields populated by coffee, insects, and farmers.

This shouldn't be surprising. Bruno Latour and Steve Woolgar's classic ethnography of scientific activities in the Salk Institute, *Laboratory Life* (1979), documents numerous careful observations of scientists refrigerating specimens, operating instruments, discussing reams of readouts from analytic machines, scribbling notes, and organizing scientific papers. After cool observation, these ethnographic researchers reach a key conclusion about what is going on: rats go in and text comes out. In between, facts emerge from complex negotiations between scientists hovering over test tubes, microscopes, sketches, and printed curves. Debate and consensus slowly self-organize as Salk Institute scientists come to agree on the meaning presented to them by the things—artifacts, machines, neural tissue, and inscriptions—they simultaneously produce and encounter in their work. At the center of discovery, therefore, is a universe of things other than the researchers themselves. This is all the more remarkable given the actual architectural theory of the Salk campus, which is precisely designed as a model laboratory, to encourage scientific interaction.

Here, conversely, art is forethought, a directing catalyst that places things in-the-world into the path of potential interacting actors and investigators.

The practical implication of this admission, that multidisciplinary groups become interdisciplinary communities primarily through their urgent and sometimes affective relationship with common things (rather than common ideas), is to worry less about socializing scientists with each other and to spend more time and resources convening conversations about specific things.

It also means that multidisciplinary engagements are most fruitful when they are united around a thing of concern. Knowing this, the role of art in collaboration becomes all the more urgent. By representing and invoking the relationships between people and objects of the world, creative expressions initiate further collaborative investigations. This stands in stark distinction to the way projects that involve art–science collaborations are more typically imagined, where creative expression is deemed useful merely to illustrate an insight or finding after the fact. Here, conversely, art is forethought, a directing catalyst that places things in-the-world into the path of potential interacting actors and investigators. Necessarily, it cannot be anticipated that every artist, writer, or researcher is precisely on the same track, or even working from the same motivation (be it aesthetic, political, or mercenary). It does mean they share a common concern: a real thing.

This means that a conversation about a dry river is one where one can reasonably expect inquiry and expression to run on multiple tracks (indeed: "parallel play") with unexpected dialogues and cross-fertilization taking hold in fragments and in unexpected chains of association. As is increasingly acknowledged in the field of genetics (Heinemann and Kurenbach 2009; Heinemann 2011), this kind of chaotic "horizontal gene transfer" between creative and scientific species is likely to cause far more mutation for selection, and so a greater role in the evolution of our ways of knowing, than the controlled, vertical, cross-bred inheritance of classical genetic thinking.

POLITICAL POSSIBILITIES OF OBJECT COMMUNITIES

And so biking past the Rillito River on a Saturday morning in September, having been soaked in the expressive and analytic engagements of a diverse scientific and artistic community, it is unsurprising that I would encounter it in a new way, and feel a strange and sudden urgency to think about its fate in a growing conurbation, choked with toxic runoff and heavy development, and ever-growing demands for water. These concerns might be racked up to my long-standing, on-and-off, commitment to "environmentalism."

But something else is at work here; the river, brought to my attention in new ways by my brief interactions with a broader interdisciplinary community of investigation, now seems a forceful voice by itself, expressing or enacting—through its bare form, snaking through the city, splitting the wealthy from the unwashed, flowing fiercely in unexpected moments—a political demand on the city itself. The river is a constant reminder of the ever-incomplete modern project to insulate the human urban world, in its appetite for inputs (e.g., water) and its necessary exports of excess (e.g., wastewater), from an uncontrollable world of nonhuman things (Moore 2009). The Rillito commands attention precisely because it announces the way the world exceeds our desire for a tidy urban island. Simultaneously, the life that cannot be contained within the river's banks, including the thousands of bats that emerge every night throughout the summer and are center stage on Bat Night each year, constantly insist on representation in our urban community. Ideas, representations, and theorizations of the river certainly do a great deal of this work, but much of it is done by the river itself, its teeming animal and plant residents, its ubiquity across the city, and its uncontainable presence in so many aspects of daily life.

As political theorist and philosopher Jane Bennett points out, "environmentalism" neither describes nor encompasses this affect or effect. Rather, she insists, theories of democracy will have to transcend the anthropocentrism that even "environmentalism" invokes (with its concern for an external nature, passively protected or destroyed by humanity) to admit the role of nonhumans as part of the broader public. Such an effort will require that we "devise new procedures, technologies, and regimes of perception that enable us to consult nonhumans more closely, or to listen and respond more carefully to their outbreaks, objections, testimonies, and propositions" (Bennett 2010: 108). In this sense, convening scientists, scholars, artists, designers, or other interdisciplinary groups around an object is not only likely to strike intellectual sparks and pave the way for new kinds of knowledge (better science), it is also likely to inspire new kinds of concern, through the recognition of the role of the object in question in participating in its surroundings and causing new outcomes (better politics). Not only does the current formal model of interdisciplinarity fail to acknowledge interdisciplinary work as it really occurs on the ground, through heterogeneous assemblages of people and things learning about one another, it also draws attention away from the political possibilities that such encounters present. By talking through a dry riverbed, rather than merely about it, it is far less difficult to hear the voice of the river itself, and to acknowledge its interests in a greater scientific and municipal polity.

REFERENCES

Bennett, Jane. 2010. *Vibrant Matter: A Political Ecology of Things*. Durham, North Carolina: Duke University Press.

Committee on Facilitating Interdisciplinary Research and Committee on Science, Engineering, and Public Policy. 2004. *Facilitating Interdisciplinary Research.* Washington, D.C.: National Academy Press.

Heinemann, Jack A. 2011. "How the world looks to a gene." Presented at Understanding and Managing Ecological Novelty, Monte Verita, Ticino, Switzerland.

Heinemann, Jack A., and Brigitta Kurenbach. 2009. "Horizontal Transfer of Genes between Microorganisms." In *The Desk Encyclopedia of Microbiology*, edited by Moselio Schaechter and Joshua Lederberg, 582–591. Oxford: Elsevier.

Latour, Bruno, and Steve Woolgar. 1979. *Laboratory Life: The Construction of Scientific Facts*. Edited by Jonas Salk. Princeton, New Jersey: Princeton University Press.

Moore, Sarah A. 2009. "The Excess of Modernity: Garbage Politics in Oaxaca, Mexico." *Professional Geographer* 61(4): 426–437.

Morton, Douglas C., Ruth S. DeFries, Yosio E. Shimabukuro, Liana O. Anderson, Egidio Arai, Fernando del Bon Espirito-Santo, Ramon Freitas, and Jeff Morisette. 2006. "Cropland expansion changes deforestation dynamics in the southern Brazilian Amazon." *Proceedings, National Academy of Sciences of the United States of America* 103(30): 14737–14641.

Perfecto, Ivette, John Vandermeer, and Angus Wright. 2009. *Nature's Matrix: Linking Agriculture, Conservation and Food Sovereignty.* London: Earthscan.

Vandermeer, John, Ivette Perfecto, and Stacy Philpott. 2010. "Ecological Complexity and Pest Control in Organic Coffee Production: Uncovering an Autonomous Ecosystem Service." *Bioscience* 60(7): 527–537.

Walsh, Stephen J., Joseph P. Messina, Carlos F. Mena, George P. Malanson, and Philip H. Page. 2008. "Complexity theory, spatial simulation models, and land use dynamics in the Northern Ecuadorian Amazon." *Geoforum* 39(2): 867–878.

Wright, Angus Lindsay, and Wendy Wolford. 2003. *To Inherit the Earth: The Landless Movement and the Struggle for a New Brazil.* Oakland, California: Food First Books.

Climatic Landscaped Habitat

JONGWOO KIM

B Arch, College of Architecture + Landscape Architecture, 2012

Advanced Architectural Design Studio: Water+

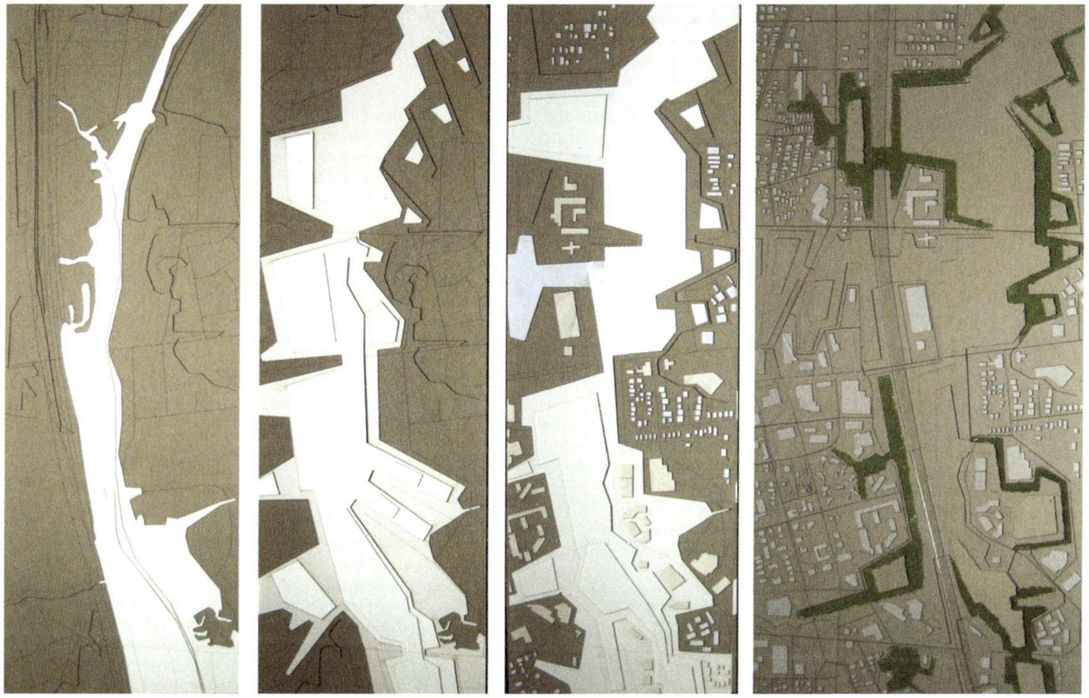

Study models exploring topographic manipulation to encourage water retention and groundwater recharge | chipboard | 22" x 32" | 2010

The current path of the Santa Cruz River flows north. The dry river collects its intermittent flow from substantial seasonal storms. The objective of this project is to create an urban-scale intervention to slow the movement of water flowing into the river. The sculpting of land allows the rainwater to collect into the designated basins and percolate into the ground. The excavated earth is molded into artificial hills, which are used as shelters for residential development to offset the loss of habitable land. Housing units are sunk into each hill, utilizing its mass for protection against the summer heat, with roof angles to allow winter sun to fill the living space. The land forms and the integration of a dynamic living condition creates an advantageous dependency upon the sun, rain, and physical topography of the site.

The Nile and Colorado Rivers

DANIELLE ALVAREZ

B Arch, College of Architecture + Landscape Architecture, 2011

Interdisciplinary Seminar: Critical Issues in Art + Design, 2010

For centuries, border-crossing rivers have dictated relationships between neighboring communities. Conversely, the inherently ephemeral dynamics of these rivers have been forcibly influenced by evolving societal factors. Declining water supply and sustained population growth have become two sources of potentially violent conflict along these rivers. Previous agreements and settlements regarding water rights have been retracted due to long-term damming and overuse of these integral arteries.

This study seeks to illustrate the issues surrounding the Nile and Colorado Rivers through the layering of statistical information pertaining to population, demographics, and geography. Envisioned as an exhibition at human scale, the design becomes a tool for education. Each element of the experience represents a statistic, and each statistic emphasizes the literal "pull" that humans have on these desert rivers. The magnitude of that pull is dependent on the amount of water flow, the length of the river, the size and population of neighboring countries or states, and the wealth of the surrounding area.

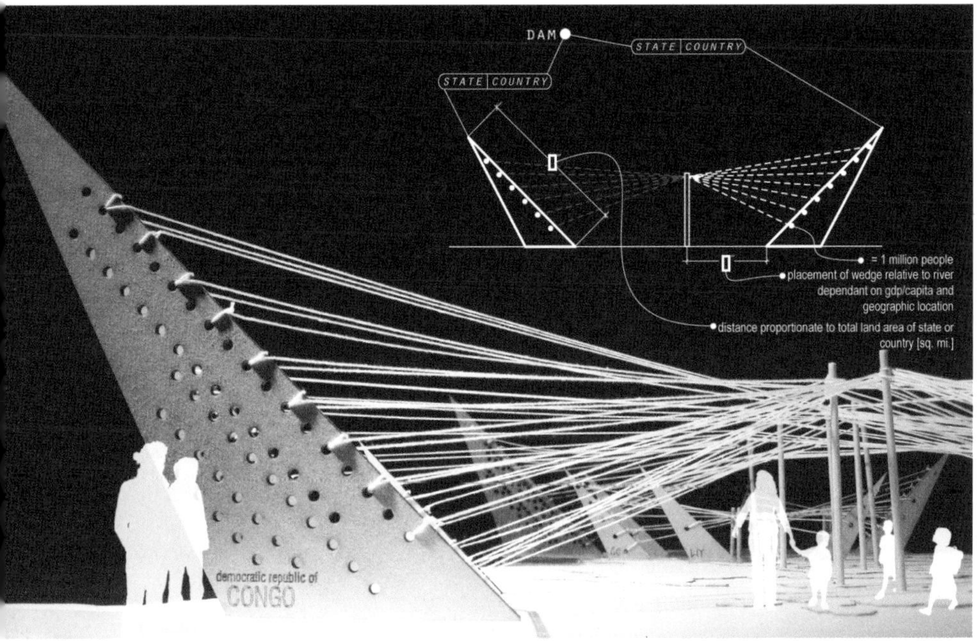
DAM
STATE | COUNTRY
STATE | COUNTRY
= 1 million people
placement of wedge relative to river dependant on gdp/capita and geographic location
distance proportionate to total land area of state or country [sq. mi.]
democratic republic of
CONGO

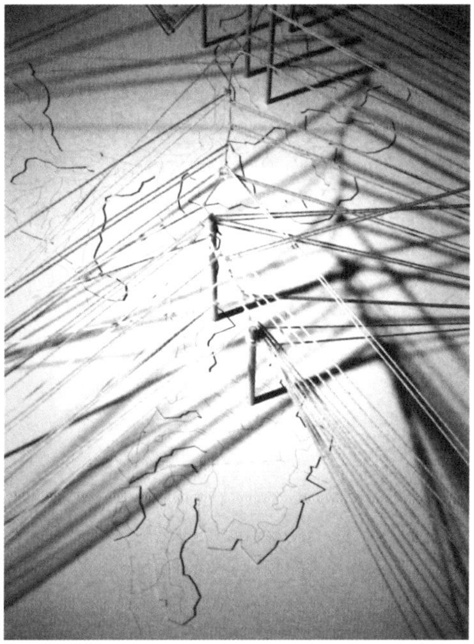

Tucson's Water Future

JEFF LEINENVEBER

BFA, Visual Communication, School of Art, 2011

Interdisciplinary Seminar: Critical Issues in Art + Design, 2010

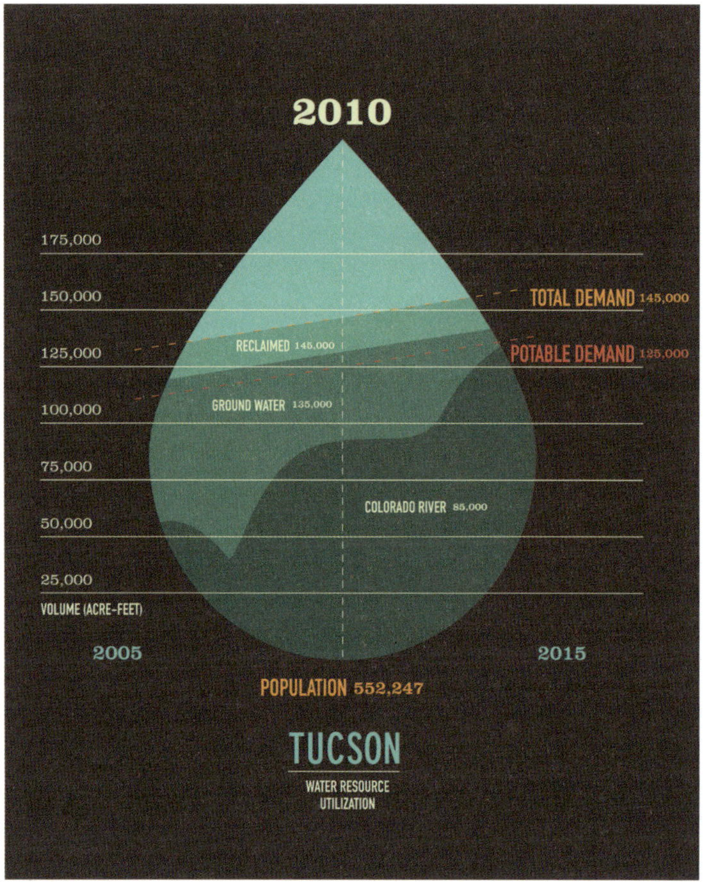

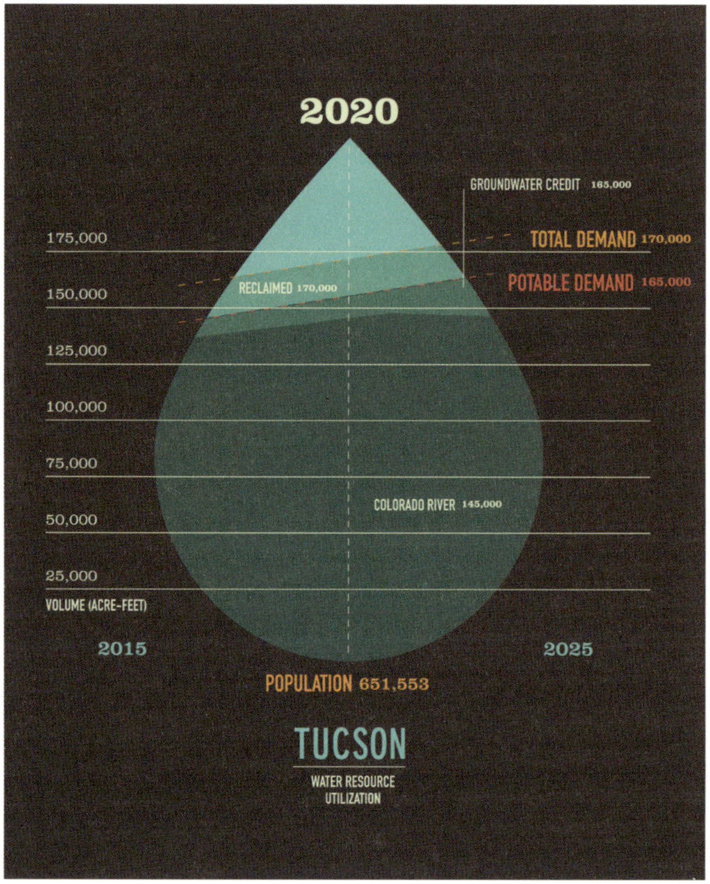

In the interdisciplinary seminar, Tucson Water's Chief Hydrologist Wally Wilson presented the city's water usage with graphs to illustrate Tucson's long-range water plan for the present and projected years to come, as well as the sources from which the water will come. The graphs also highlighted "the gap" between projected demand and supply, and how Tucson Water plans to fill the gap in the coming years.

My goal was to research, analyze, and organize the data necessary to communicate this information in a more simple, straightforward, and aesthetic way to clearly convey the very complex subject of Tucson's water. This series I designed was given to Wally Wilson for use in his presentations to help better educate the public on the importance and future of water in Tucson.

COMMENT FROM WALLY WILSON

The shortfall in supply projected in later years can be addressed in a number of ways: additional conservation, a slower rate of population growth, fully utilizing the recycled water the city is entitled to, leasing of agricultural or Indian Nation CAP allocations, and lastly pumping the finite amount of groundwater that the city can legally pump.

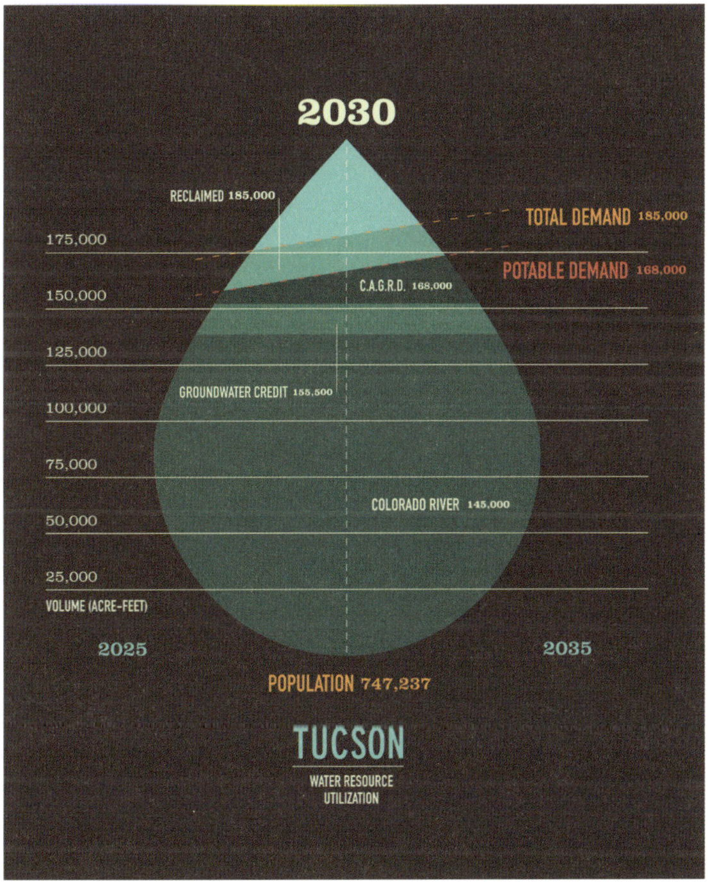

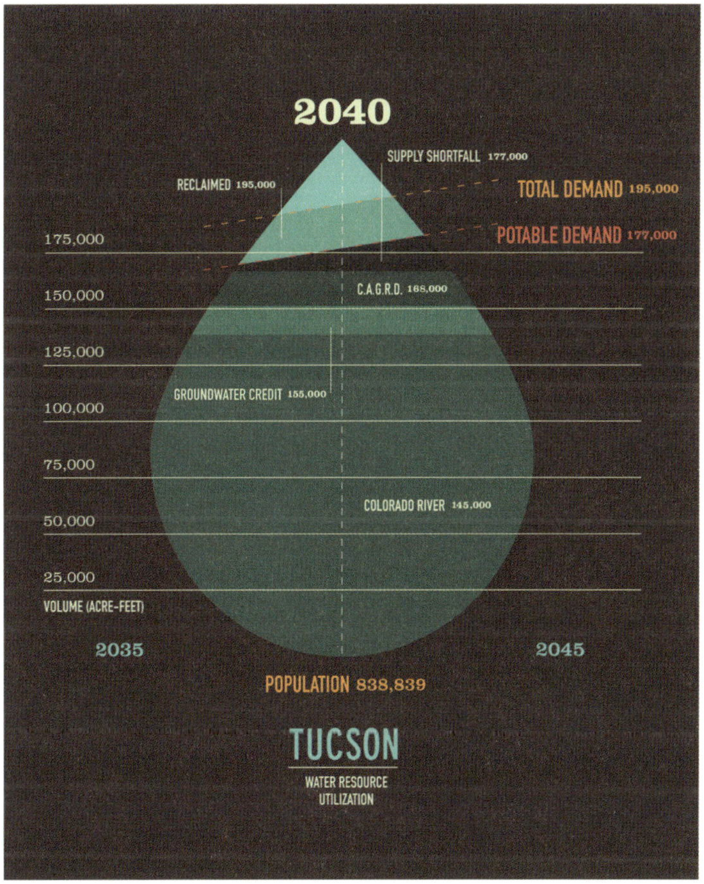

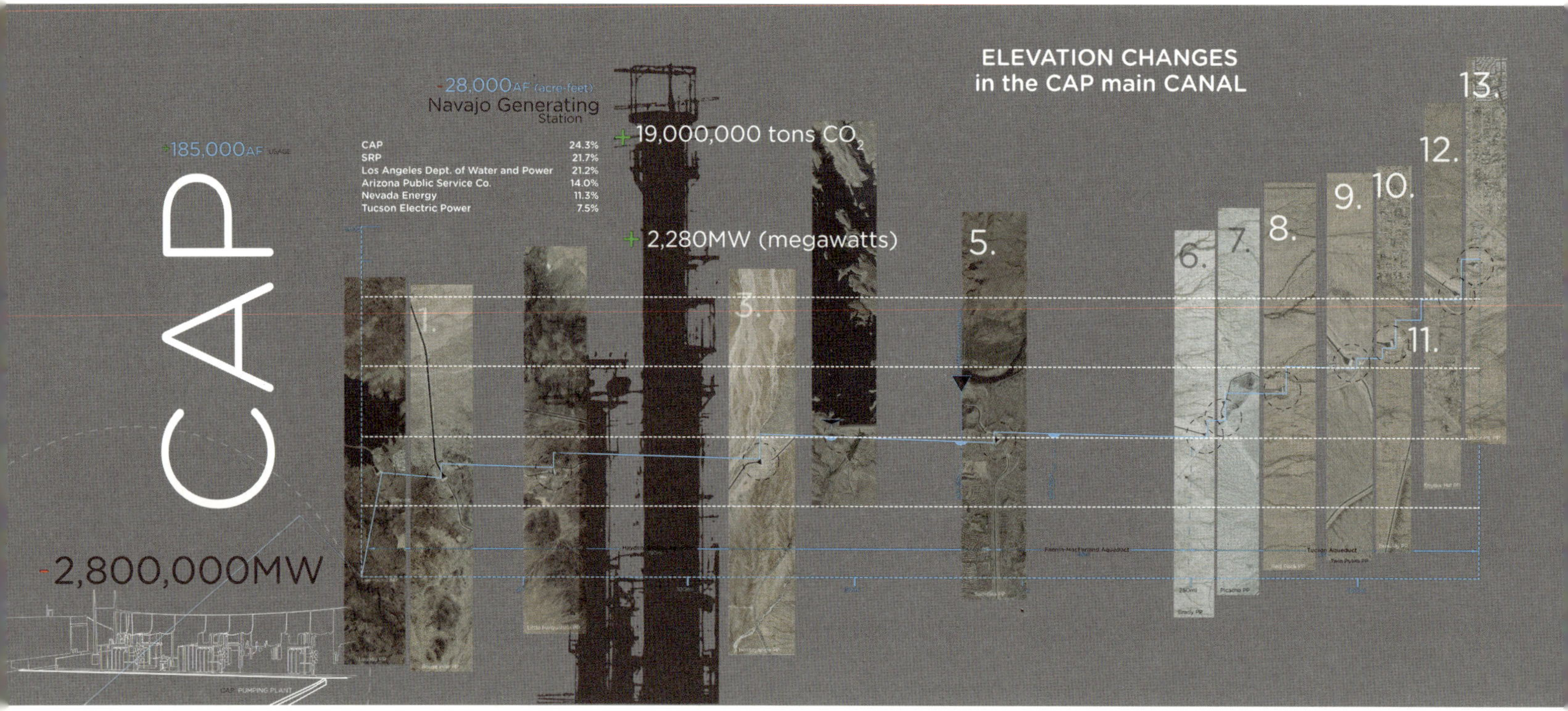

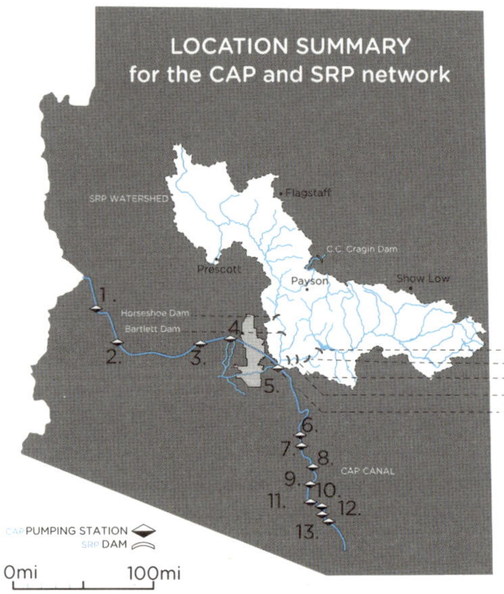

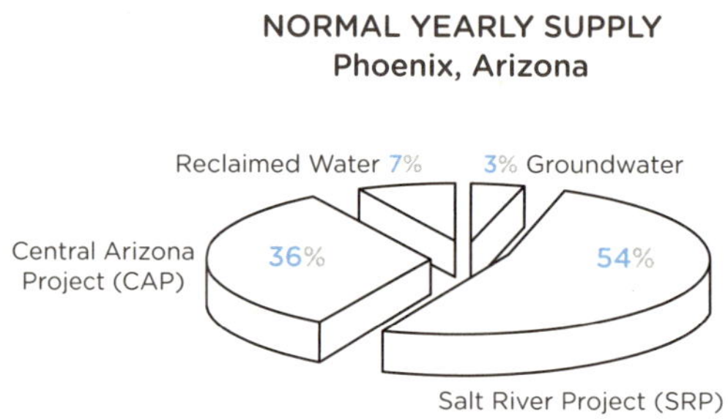

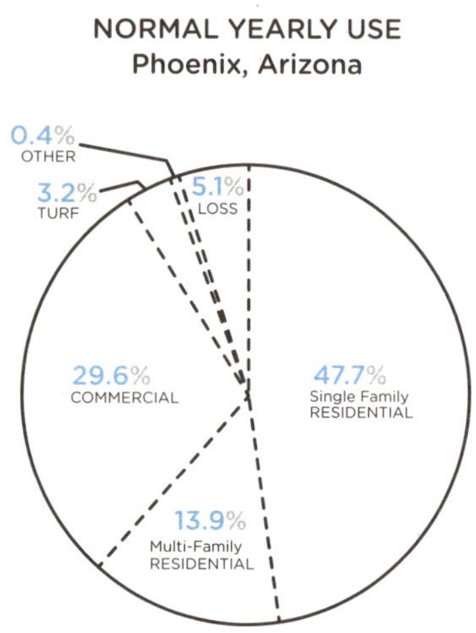

ABOVE TOP **Illustration of the interconnection of water conveyance and energy use along the CAP, including the vertical feet pumped and horizontal distance travelled from the Colorado River to Tucson.**

LEFT **SRP watershed with its hydroelectric dams and the CAP canal with its pumping stations.**

ABOVE AND RIGHT **Water sources supplying Phoenix, Arizona, and distribution by types of use.**

Tucson + Production

MATHEW PROPST

B Arch, College of Architecture + Landscape Architecture, 2012

Advanced Architectural Design Studio: Water+

Tucson + Production is a proposal based on semester-long research about local/global water issues, applied to Tucson conditions projected for the year 2050.

The initial phase of this research focused on water issues in my home city of Phoenix, Arizona. Nearly half of Phoenix's water supply comes from protected Salt River Project (SRP) watershed lands in the mountainous terrain just north of Phoenix, while another 43% comes from the Central Arizona Project (CAP). By contrast, Tucson, plus several tribal, agricultural, and other areas, rely very heavily on the CAP.

CAP is a 336-mile system of aqueducts, tunnels, pumping stations, and pipelines that carry water from the Colorado River to central and southern Arizona and Tucson. CAP was completed to Phoenix in 1986, and to Tucson in 1992. This system is deeply problematic in that the water it carries must travel vertically to an elevation of more than 2,600 feet above the starting point, requiring a massive amount of energy. Utilizing a series of pumps and an integral storage reservoir (Lake Pleasant) on the Agua Fria River, the canal was designed to convey CAP's 1.5 million acre-foot (AF) entitlement, and is capable of carrying up to 1.8 million AF per year. The City of Phoenix has access to approximately 185,000 AF of CAP water.

The project proposes that future architectures stop using the CAP's costly and inefficiently transported water supply and make use of water that can be found locally. Contingent upon increasing current conservation, harvesting, and recharge efforts to restore Tucson ground water, the project *Tucson + Production* assumes that the Tucson Basin aquifer water levels will be significantly higher in 2050 than today. The proposed project is sited on the southern bank of the currently dry Rillito—a decision based upon pre-1860 water table levels, information regarding this area's hydrologic soil group, and the presence of shallow groundwater. By 2050, higher groundwater levels would reinvigorate the entire riverbed where water and vegetation would once again become a normal occurrence.

Tucson + Production proposes a building typology for the local production of food and energy. Using conventional soil-based farming methods, farming just a half-mile area along the Rillito River would satisfy half of Tucson's population. Introducing new methods of farming such as hydroponics, in combination with conventional farming, would not only feed all of the Tucson area but also create a food surplus that could be sold elsewhere, all while using only the local water supply. The proposed building type incorporates a mixed program, including retail, industrial, and residential uses. This creates a work/live setting in which people can work in a wide variety of jobs, ranging from high-tech (solar, wind, hydroponics), to labor (farming, repair), to retail, all located within this typology.

(A) The Rillito and its tributary washes and creeks (white), existing riparian areas (green), and the 0.5 x 18-mile proposed agricultural strip (red tinted area). (B) Project study area.

THE AVERAGE PERSON CONSUMES 6.6 POUNDS OF FOOD A DAY. THIS EQUATES TO APPROXIMATELY 2,200 POUNDS A YEAR REQUIRING 26,910 FEET OF FARM LAND, INCLUDING FEED FOR ANIMALS, TO MEET THIS DEMAND.

THE PROJECTED 2050 POPULATION OF PIMA COUNTY IS EXPECTED TO EXCEED 1.8 MILLION PEOPLE. THE AREA OF LAND NEEDED TO ACCOMMODATE THIS POPULATION, ACCORDING TO THE AVERAGE LAND-TO-PERSON RATIO, WOULD BE APPROXIMATELY 24 SQUARE MILES OF LAND.

2,200 pounds/year　26,910 feet　1.9 million people　24 square miles

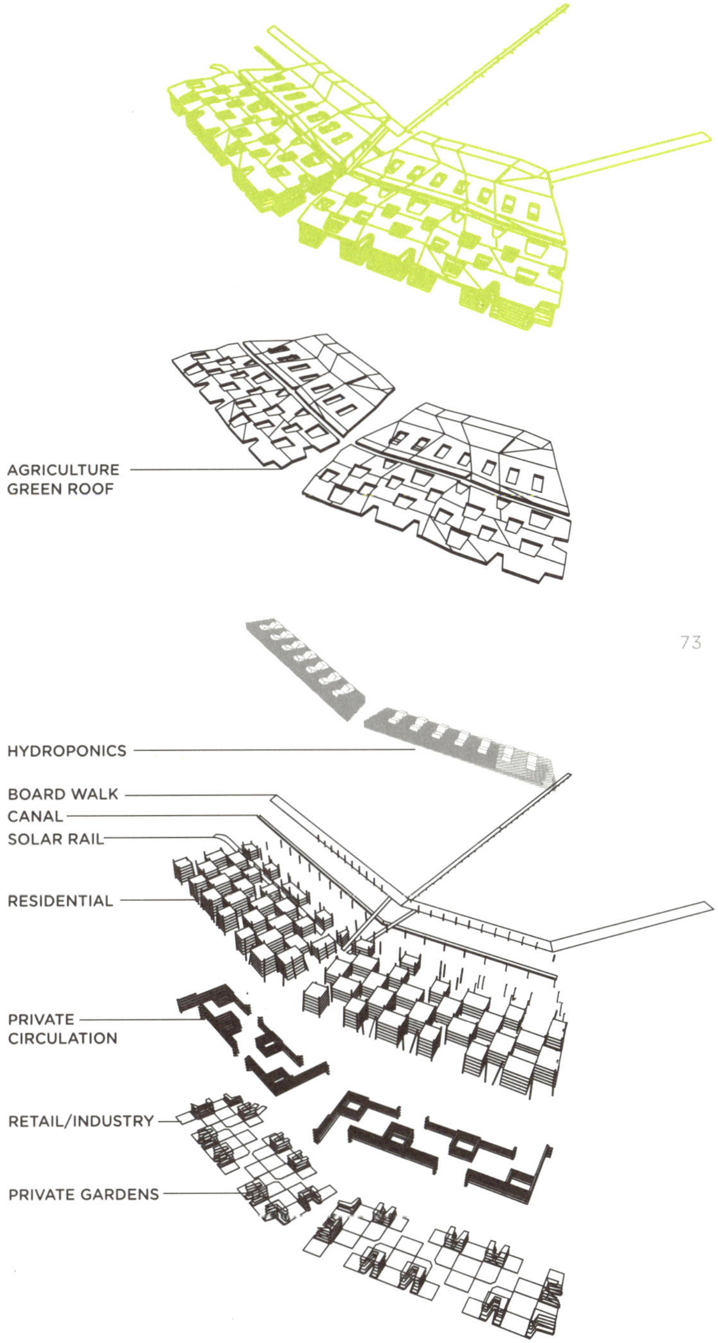

ABOVE TOP Rendered views of greenhouses and roof from river.

ABOVE BOTTOM Rendered view of residence.

RIGHT Exploded view of programs and spaces.

Industrial Receptacles

SULAIMAN ALOTHMAN

B Arch, College of Architecture + Landscape Architecture, 2012

Advanced Architectural Design Studio: Water+

Agriculture was Tucson's predominant activity in the 1940s, constituting 84% of its economy. However, it decreased to 73% by 1951 and remained constant up through the 1970s and 80s, when government acquisition of farmlands made way for urban development in response to the region's population growth.

As part of a larger movement begun in the United States in the 1980s, Tucson founded its Community Supported Agriculture (CSA) system in 2004 to encourage local agriculture and deliver a variety of crops to the local community. As a networked system of agriculture linking local farms, people, and communities, CSAs also foster a high level of social interaction lacking in the big-box food systems of today.

New businesses in Tucson's industrial district demand water as a resource for their processes, generally related to building construction and fabrication trades. The sharp increase in water demand—generally due to urban population growth, but contributed to by high-tech and fabrication industries and the increased popularity of locally grown produce—contributes to the region's water scarcity. Creating a loop for filtration and reuse, between industry and agriculture, will ease the demand for water.

The potential to remediate Tucson's industrial wastewater led to the proposal of *Industrial Receptacles*. The proposed system of receptacles is an intervention into open spaces within the existing blocks of Tucson's industrial area. The form of the receptacles is adaptable to the space between industrial buildings, with each receptacle having an outer and inner shell. The outer shell houses industrial wastewater micro-treatment plants. These separate and filter wastewater into plant nutrients and reusable water; the nutrients and water are then pumped to research labs within the shell for quality check and refinement. Finally, these are pumped to tanks that supply aeroponic growing fields located in the upper layer of the outer shell. Once harvesting occurs, crops are delivered to a market occupying the inner shell. Thus, the cycle of water filtration, plant growth, harvesting, and distribution encourages human interaction and raises the awareness of the importance of wastewater as a resource.

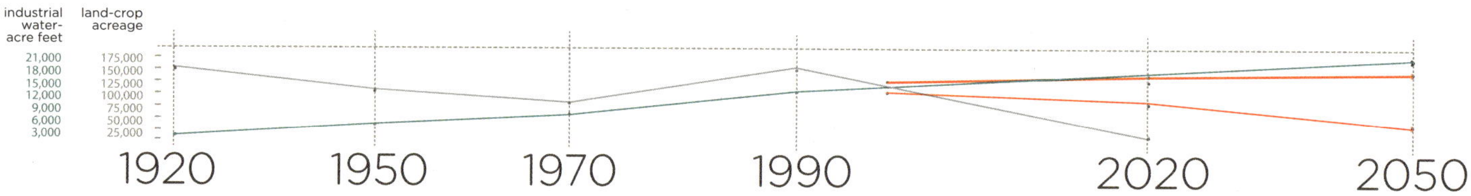

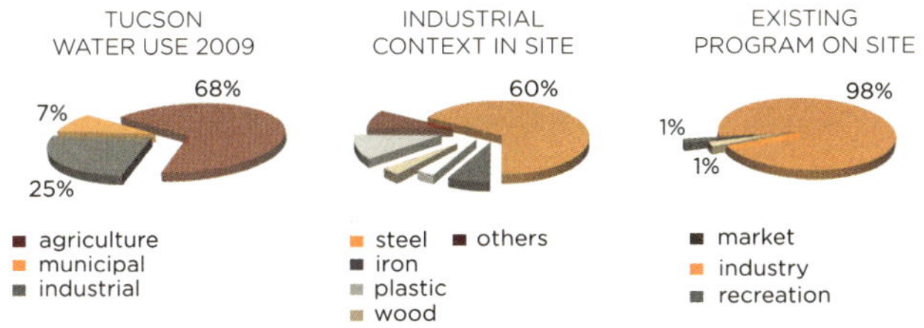

OPPOSITE PAGE Timeline charting Tucson agricultural activity and water use.

TOP LEFT Tucson water use in general, and in the industrial area.

BELOW LEFT Industrial wastewater-to-farming loop.

BELOW RIGHT Mineral content of groundwater, wastewater, and as required for aeroponic farming.

BOTTOM Water-agriculture-market connection.

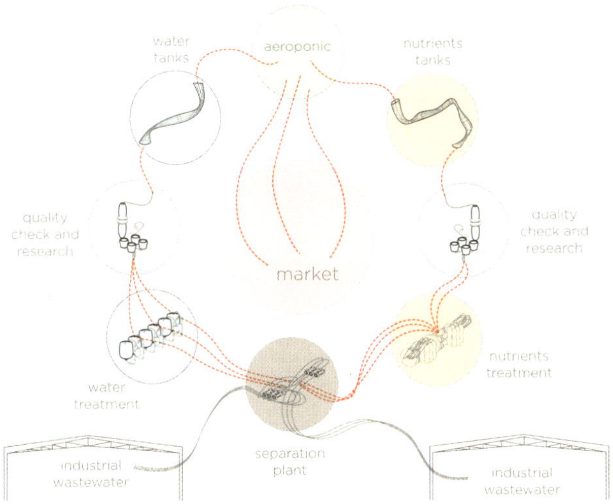

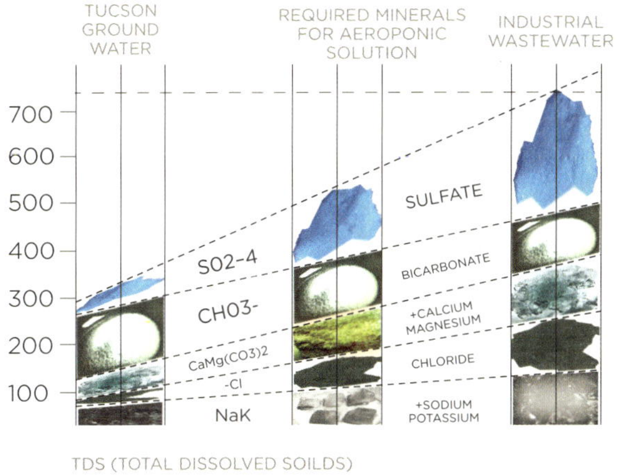

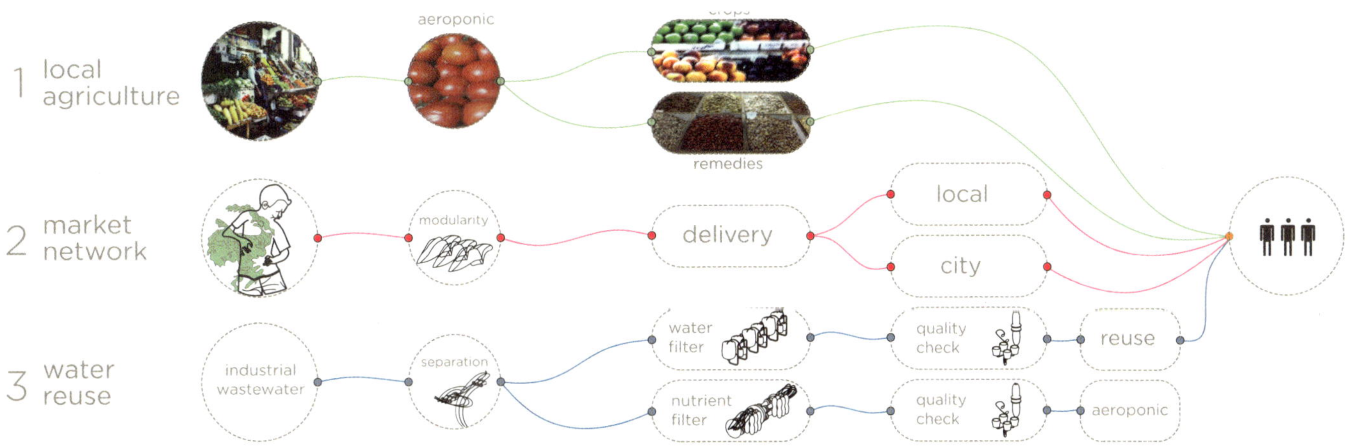

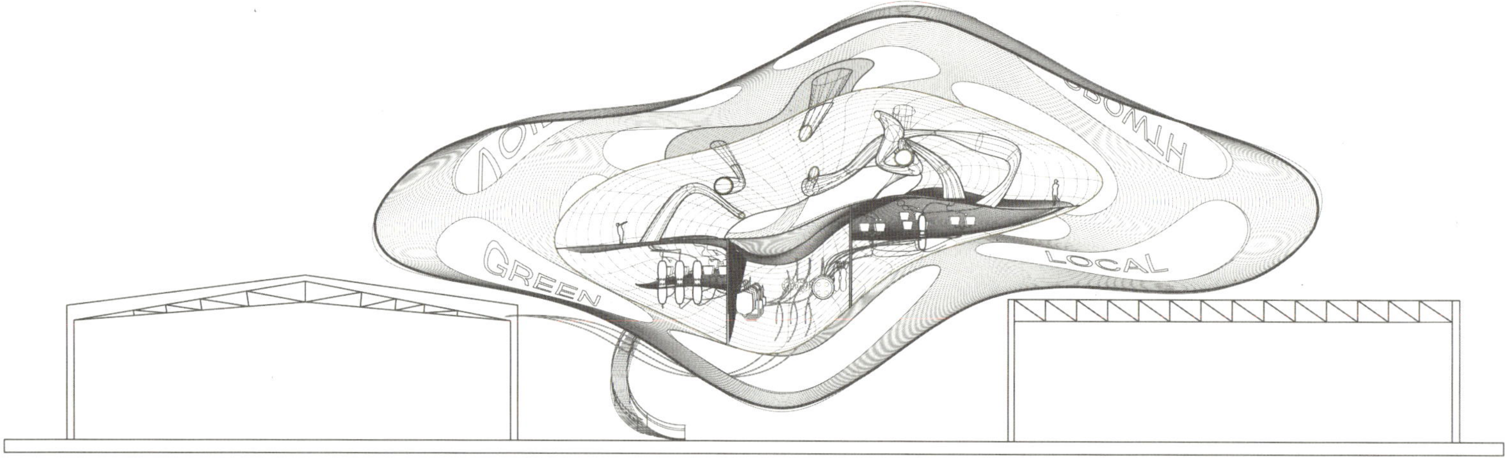

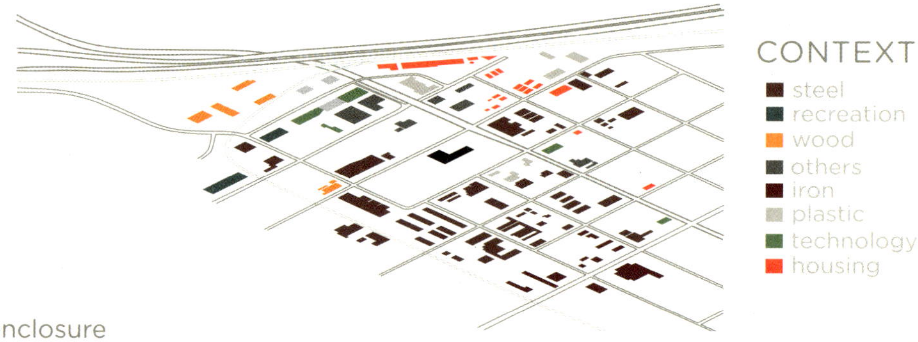

ABOVE TOP **Transverse section through elevated receptacle and existing industrial warehouses.**

ABOVE MIDDLE **Industrial area context, with vast open spaces.**

ABOVE BOTTOM **Receptacle morphology in relation to industrial area open space.**

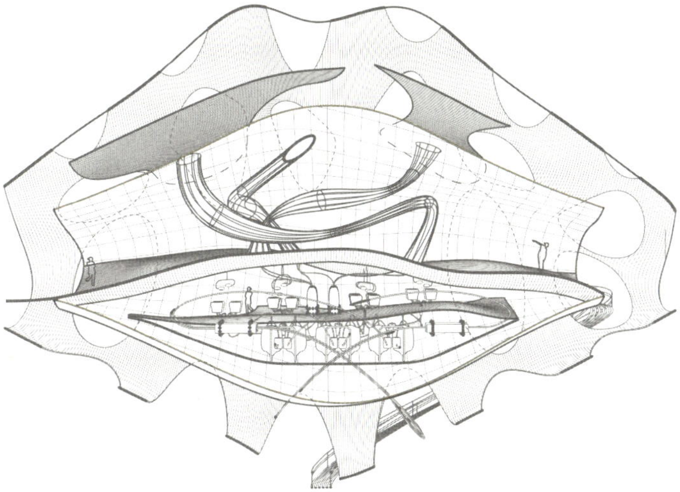

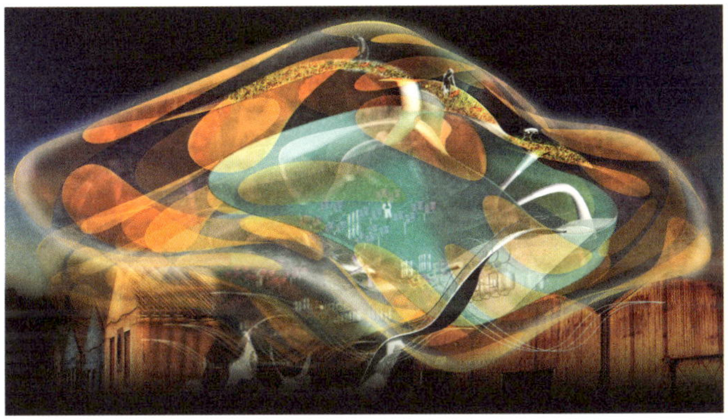

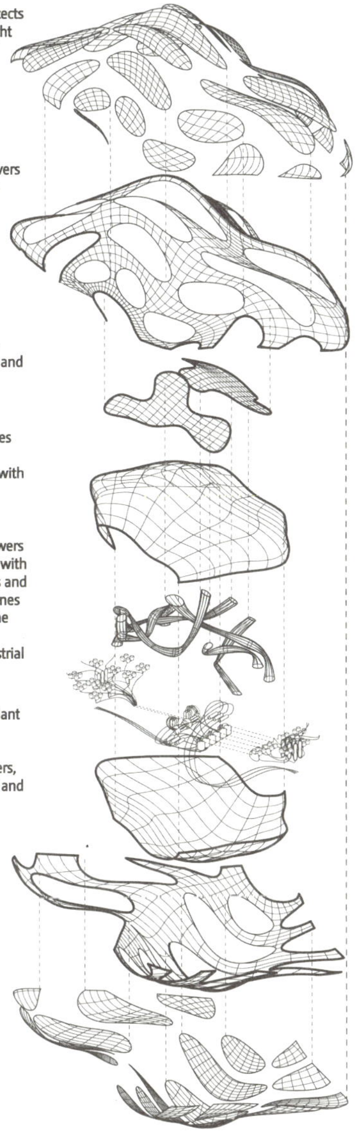

LEFT TOP Longitudinal section through one receptacle, with water filtering, lab, and greenhouse.

LEFT MIDDLE Exterior view of the industrial receptacles against warehouses.

LEFT BOTTOM Market interior.

ABOVE Blow-apart diagram of the receptacle inner and outer shells.

ANDER

The channel at right surrounding—disappearing into—the spine of the book is the gutter. In the unlikely event of a water landing and your book being submerged, water would collect in the gutter. The pages would soak through and fall apart given time. The glue would dissolve. The stitching would rot like a long-submerged body in a river or a lake. By this point either you would have survived or you'd be falling apart. The binding would then fall into its component pieces, ruining the illusion of the artifact's wholeness you're holding. What is in your hands is just paper, collected, printed on. What you are holding is just ideas and language—the usual interlocking systems of signs—printed on paper. It is not engineered to channel or hold water, but it can do so in a pinch.

This year natural disasters have been everywhere, it seems. As I write this, "Tornadoes kill 6 in Oklahoma, Kansas" (CNN), and we're less than a week past the devastating tornado in Joplin, Missouri, not to mention the more obvious man-aided nuclear disasters occurring after the epic tsunami in Japan, the tornado system that leveled both of the places I lived in Tuscaloosa, Alabama, from the map a couple weeks back, then the crazy freeze this winter in Tucson that killed many of the saguaros though they won't keel over for some years, their death delayed, and then the massive floods in New Orleans, again, the Red River in North Dakota, the eruption of Grimsvotn, the Icelandic volcano whose ash cloud drift threatens to scuttle my summer travel. My mother-in-law is freaked out about the end of the world prediction that didn't come to pass a couple weeks ago on May 21st, 2011, not to mention the Mayan apocalypse predicted for next year, just to name a few anxious highlights. I mean to say that the world's tribulations are present like weather in my mind, this month, going into the Tucson monsoon season.

in River

MONSON

The Rillito River does not channel water through semi-arid Tucson much of the year. It once did, and like many modern things—like the trains in Upper Michigan, my homeland; like Western Union's telegram service; like the pinching of my childhood cheek by tottering relatives; like the streetcars in Grand Rapids, Michigan, or most other American cities; like the increasingly anachronistic telephone cables strung all along our highways into the second-largest engineering project after the interstate system; like the blood through the bodies of our forebears, our betters, our mothers, our fathers, now dead; like the language recognition centers in my brain that fail now to bring the words I want to mind or tongue at awkward moments; like the awesome chugging rhetorical power of the periodic sentence that cannot go on forever even as it aspires to—the river channels water no longer, not year-round. But the channel's there, isn't it, plain as a wide June day, scrubbed clean in many parts and bristling with brush and trash in others, hundreds of feet across at points, hemmed in mostly by concrete culverts, effectively separating the city from the ritzier homes and restaurants that populate the foothills. It separates via space and threat of sudden flood surge, a reminder of the force of water when it's here, a reminder of the force we deploy in trying to contain it, a reminder of the risk we take in ignoring it.

I have recently started running much of the year myself. I do not run high like the river or cheerful, smelly Arizona stoners. I don't know what drives my spur to run: a fight against my own personal beltway westward expansion, perhaps. A connection to a fitter vision, finer version of myself. A bit of personal engineering: clawing some control away from age and the natural tendency of the body.

Any individual one of the brain's channels—axon, neuron, dendrite—is idle much of the time, though not to say empty or dry, exactly. After injury or catastrophic change the channels reconfigure, find new ways to get the signal through the network. (As aficionados of weird brain injuries know, sometimes they don't redirect so successfully.) Perhaps like desert denizens they lay where it's cool and await the occasional electrical impulse, at which point they are signaled, woken and singing, alive.

The writers had to engineer a plot into which they could channel the songs.

I imagine most musical theatre is built this way—songs first, plot second, with some backfill as necessary. One structure is built to accommodate—contain—control—another. That might account for the odd plots of Gilbert & Sullivan, for instance.

It's fair to say that my life is suspended between two poles—the wet or snowbound landscapes of Michigan that still shape the way I think and dream and write, and the hazy desert loneliness of Arizona, where I now live and work.

And when the water's here in the monsoon months of July and August, it's everywhere: like nothing in the north except maybe a blizzard surge or avalanche. The Rillito runs high when it storms, meeting and joining the Santa Cruz and moving on north, bringing what benefits water brings to the arid bits of West that it reaches less and less each year.

It is therefore possible to execute a water landing in Tucson at least a dozen mostly monsoon days a year.

What else runs: my car (though not my old, sold, or crashed cars); my computer; my animals when playful or fearful; machines; programmed processes in machines; my nose in the dry air of the desert; coyotes, rabbits, dogs, humans, lizards, and rare javelina through the wash in the mornings when I run and spark their chase or flight; the blood through the veins; the thoughts through the brain late at night when they should be shutting down into easy, ecstatic dreams.

Trivia: The script for the movie musical *Singing in the Rain* (1952) was only written after the songs were already in place.

Back in Michigan, I'm seventeen and dancing jigs poorly as a sailor in *H.M.S. Pinafore*, Gilbert & Sullivan's sea-set piece of musical theater as it rains outside again: you can hear it on the roof. In Michigan water's abundance is an apparently incontrovertible fact. What is Michigan if not a wet state, a water state? Bordered by four of the five great lakes, Michigan's vision of itself as a vacationland is no less dependent on water than are the cities of the arid west. Fishing, boating, beaches, tourism, even skiing, hunting, and snowmobile riding—these are all built on water. Everything is lush, green, bushy, buggy, muggy, prickled, itchy, and wet except when the state is blanketed by the yearly onslaught of snow (still water, we're reminded, when our frozen boots melt on return in the house). Michigan is all coast, two peninsulas separated and defined by water, settled by water, driven by water, separated from Canada by water.

It's fair to say I didn't think much about water before moving to Arizona, where water means absence, means scarcity, means need, means conservation, except when it monsoons, and the river runs high.

For those who don't consider the Rillito or Santa Cruz to be rivers because they don't run year round, these occasional spectacular events serve as useful and painful correctives to their notion of *river*.

Much human engineering works this way: we design our structures only up to specific breaking points, knowing they'll be eventually exceeded. The key is knowing when they fail.

Admittedly the expense of anything engineered to be asymptotically approaching permanence or perfect safety would be unfeasible.

The only flawless song is a funeral song, chosen to fit the form of the funeral.

So which is the river, we ask: the channel itself, the water that flows through it?

Some constraints aren't mutated or formed over time, but designed. For instance, the pagebreak

Alternating landscapes like this, if done quickly enough, produces a double image, a composite of the two that compresses the space between them.

The channel through which the river occasionally runs high is partly natural and partly man-made. In the natural corner, the river's carved this space out from the earth over thousands of years. In the man-made corner, its course is constantly monitored and engineered, rejigged, tweaked, shunted, diverted, plotted, adjusted, and planned. In 2010 the city of Tucson dredged out and buttressed the channel walls of the Pantano wash on the east side of the city which I run many mornings. The Pantano feeds into the Rillito, which feeds into the Santa Cruz, and its watershed stretches from the Cienega Creek to the Colossal Cave out toward the Chiracahua Mountains.

Four years earlier the concrete bank protection along an executive park and a townhome development (our modern names as ugly as the structures they describe) was damaged in a 35-year flood event (a 35-year flood event happens roughly every 35 years, and has an approximately 3% chance of happening any given year), when upward of 15,900 cubic feet per second of floodwater passed under the Broadway bridge.

Further northwest, the Cortaro Road bridge over the Santa Cruz was destroyed completely in the flood of October 1983 and was rebuilt.

And when our structures fail, they fail big.

The only perfect safety is that of stasis, that of the virtual, that of math, attenuation then cessation of breath, that of death.

The upshot is that our engineered world is not built to last without constant intervention. So we monitor the river, tweak the river.

This isn't an idle line of thought. A great deal of money and development and, on the flipside, conservation effort depends on how we want to parse the definition of *river*.

Definitions change over time. They're flexible, descriptive, liquid. Perhaps they should be geographical too; they shift and mutate as we do. After all, we don't all agree what we mean by *river*. We don't really seem to agree what we mean by anything.

is a designed absence, as is the gutterspace. I don't think about inhabiting it often, unlike margins, which lend themselves to annotation, to notes on the marginal, aside from the text. Margins are for our fingers, so we don't obscure type with our hands, so our designed page can have a neat boundary in which it is contained. We like containment. We like order.

The whitespace makes the text an island. Perhaps to conserve the imaginative, associative space around a text, principles of classical typography eschew using this space except as caesura.

But as with any apparent absence, this preserve is soon inhabited.

Is this list of absence an absence

A pause

It's hard not to be aware
It doesn't take much, though, to forget it again, since there is still so much world with which to contend daily. We're used to forgetting. We can't not forget. Our brains can't hold everything in memory, hence that whitespace around our ideas, our memories—anything could connect to anything, could be summoned by an image, redirected by a scent. So we internalize the absence of things as natural.

It's hard not to be aware
though our brains forget it when reading. How often do you think oh, the gutter, okay, this page is done, so what do I do next? How often do you think yes, left to right, top to bottom in the western world, jump across the gutter in the center of the spread and onto the next block?

You're not allowed to build in the preserve that cordons off (and protects (and contains (perhaps like parentheses))) whatever we define as the Rillito River.

In nights and mornings wildlife fills this absent river, moving silently through the city. Much of that wildlife is now rare or missing, not seen for decades. The following long list of species hasn't been seen in the Pantano Wash since the dates in parentheses: Northern Goshawk (1999), Arizona Giant Skipper (1988), Poling's Giant Skipper (1967), Gila Longfin Dace (2004), Baird's Sparrow (2003), Felder's Orange Tip (1992), Sabino Canyon Damselfly (1988), Giant Spotted Whiptail (2005), Western Burrowing Owl (2004), Northern Gray Hawk (1994), Common Black-Hawk (1977), Arizona Metalmark (1993), Mexican Long-Tongued Bat (2005), Western Yellow-billed Cuckoo (2002), Pale Townsend's Big-eared Bat (1986), Arizona Ridge-Nosed Rattlesnake (1999), Northern Buff-breasted Flycatcher (2000), Greater Western Bonneted Bat (2005), American Peregrine Falcon (2005), Gila Chub (2004), Cactus Ferruginous Pygmy-Owl (1999), Sonoran Desert Tortoise (2004), Western Black Kingsnake (2002), Western Red Bat (2005), Lesser Long-Nosed Bat (2004), Obsolete Viceroy Butterfly (1966), California Leaf-Nosed Bat (1986), Arizona Myotis (1992), Cave Myotis (2001), Pocketed Free-tailed Bat (2005), Big Free-Tailed Bat (2003), Texas Horned Lizard (1995), Gila Topminnow (2004), Chiricahua Leopard Frog (2005), Lowland Leopard Frog (2005), Yellow-nosed Cotton Rat (2003), Mexican Spotted Owl (2004), and the Northern Mexican Gartersnake (2003).

or a presence?

or a permanence?

of the absence once you see its extent, once you remember what was here before.

of the absence of text through the gutter,

Her gutter separates the verso from recto, west from the east, our space from another's, sunset from sunrise, the foothills from the city of Tucson, St. Louis from East St. Louis, the west side from the east side of Grand Rapids, Michigan; Iowa from Illinois; Missouri from Illinois; liquid from solid from vapor; the cold from the not; the haves from the nots; the darker-skinned from the lighter; Midwesterners from Southerners; this list could go on as long as our culture. Rivers are easy geographical borders, a shared feature, not part of what it divides. Even when dry, those borders remain. The split between two contiguous spaces remains. The brain sees a line.

Keep this in mind for the quiz; we'll see what knowledge your spongebrain retains.

Thus the thalweg of an essay is where the deepest and most indelible remaining thought resides when the rest evaporates or is washed or erased or edited away.

And anyone who knows anything about rivers knows that left to themselves, rivers will often
(if not always) meander

or swings back over to another curve. Or we think that's how it works. In an essay, "Meander," that appeared in the very first issue of the journal *Creative Nonfiction,* itself named for the recent American term to offer some form to contain the formless, Mary Paumier Jones suggests that, like the river, ocean currents also meander, as does the jetstream. That flow itself seems to be best modeled not as straight but as swerve, at least under certain conditions. That the river's natural swerving is not

Some information can be held within the thinking's momentum.

In the course of thinking the brain wanders, unguarded.

Minnesota recently enacted a similar law, billing stranded drivers who circumvent roadblocks to enter snow-closed highways for their rescue.

We learn that prose, like water, flows to accommodate whatever space it's given. The brain's pathways accommodate this understanding. The designer knows there is a visceral pleasure in reflowing prose blocks in page design software like InDesign because she sets up channels and adjusts the text box, watches the fluid dynamics work their magic. Today she is an engineer. Today she is a small, beautiful god.

Flash Card 1: the path of greatest depth along a river channel is called the *thalweg.*

Flash Card 2: "The sinuosity of a channel is defined as the ratio between the thalweg length and the down-valley distance" *(Arizona Dept of Water Resources Design Manual for Engineering Analysis of Fluvial Systems, 1985).*

So even if the riverbed isn't filled year-round, the thalweg is the most likely part to have water, whether runoff from precipitation on the mountains or from sudden storm.

For the runner the thalweg is likely slow going, the softest sand in the deepest channel.

in a series of sharp, snakelike, alliterative s-curves through the earth as the force of a turn eats further away at the bank; this increases until it's unsupportable, and it detaches into an oxbow

dissimilar to the way that the brain cortex is bent and looped, folded up on itself so as to fit more brain in a small space. And that the essay—best simulation of the brain on the page we've got—its natural motion, like the river's, is meander.

Some detaches, forms pools of its own.

For instance, those caught in flooding washes are subject to rescue via Arizona's "Stupid Motorist Law" which maintains that anyone rescued after driving into a barricaded wash will be billed for the cost of the rescue. You'd be surprised how often this happens. The television news takes a particular sort of glee in the footage.

The voices recorded decades ago from now-dead television laugh track audiences?

The porn actors or actresses remembered from your teenage years?

It's the mind. It goes where it goes.

For the hacker it's hard not to stray into the gutter, the deepest and often weakest point in a system. For the graffiti artist the blank is a provocation. For the oversharer. The awkward conversationalist who feares silence. For the gossip. The linguist, who, at cocktail parties, regales us with trivia: "gutter, a watercourse, natural or artificial; a small brook or channel"—obsolete (*OED*). Be careful of linguists at parties.

Perhaps you're going to the wrong parties?

We take them for data. We take them for aesthetics. Found data can be surprisingly beautiful as an aesthetic object, line on line of information spilling down the page; devoid of discernible content the form is more apparent.

As is a line of thinking—

or your preferred infinity—

As is a well-oiled hinge placed so as to pivot from one thought to another.

The lowest point where it all collects is often the best source to understand a culture.

They're fragment, suggestive of the whole.

It's connected to the dungeon—

Whether it's the trunk systems of Michigan Bell tributarying into abstract loops—

There are other gutters, of course, other abscesses, absences, holes in the ground, other ghosts that haunt these spaces.

How about the bodies in footage we conjure at night or in unguarded moments in erotic reverie that are impossible to stop?

The thought of dead lovers, gone lovers.

You sir, get your mind out of the gutter! You too, miss.

It sounds like you're talking about river, or prose.

Archaeologists at parties tell me that the gutter is where we dump our whatever.

The gutter is where telephone phreaks (I include myself) go through trash piles overflowing out of dumpsters out back of Michigan Bell buildings in search of technical papers, schematics, and streams of data uttered on paper forever.

Those multimeters with their leads! The plastic caress of lineman's handsets! Equipment is aesthetic too.

as is forever—

as is a river.

Pay attention to the absence. You can find a lot in the gutter, in the places where we dump what we disavow.

So those pornographic magazines you found down by the worksite in winter?

or the modern dungeon, the prison—where we lock what we most hate about ourselves away, where we'll bury Jared Lee Loughner.

or the telephone lines looping my dorm room of a life ago to the telephone system so that others could dial in to my private digital meander and spend time there—

or maybe we're talking about a way to cross-section a culture and see what is inside it, that inside that it prefers not to name except to itself.

Which is one good reason to keep your brain in the gutter.

An emptiness, a form waiting to be filled.

Waiting to be poured.

Pored over.

Or maybe just ignored.

A gutter is a space
a snapped taboo, a margin

As you may have heard, Tucson is on the margin of the margin of the country, within a finger's width of the border according to your map. Living here you start to realize what it is to live up against another country, living so far outside of the main tributary of the culture, what it is to border a sort of nothingness.

I don't mean this entirely as critique: it's hard to pay attention to everything. Our lives are incredibly complicated, layered, looped, processed, remixed, and filtered in ways we're barely aware of, and we can barely summon the energy to investigate some days, much less understand.

Odd to know the riverbed as a bed for the river, the sleeping anthopomorphized beast.

In Northern climes I've gone through the ice several times in waking life, and nightly in dreams.

It can be breathtaking.

The movement at the center of anything—a culture, a community—is seen best from the margin.

That gutter: a blank.

Waiting to be born into.

Waiting to be porned.

Obsessed.

elided, a kind of violence,
broken, wood suddenly unsplit.

That the border is porous is a commonplace, but it's an obvious one when you live here. Wall it up, man it up, dam it off—this accomplishes nothing. Like water, desire and imagination flow where they flow.

Of course there's something on the other side of the border (and often inside our own border), but for much of the United States it might as well be nothing, a generality, other, ether, blank spot on the map and certainly on our tax returns and drug policies, until someone legislates something crazy and everyone freaks out.

Odd to think you can understand a country without knowing its rivers, its edges.

Bodies are found in rivers all the time. In Winona, Minnesota, for instance, drunk college students fall into the river with surprising regularity. Sometimes they die. My friend Paul once nearly drowned in an Illinois river thinking that, like Jesus he could walk on water because he did once on ice, in winter.

It's terrifying and exhilarating going through a surface, breaking what you formerly thought solid.

It suggests that the rules on which we base our understanding of the world—those givens, données, paradigms, unshakable foundational truths—the walls on which we built the space that channels the brain, that tells stories to the self about the self—might just be hollow.

It can be nauseating as we rebalance, reconfigure the pathways of the brain, reckon with shock.

May their failure sustain us, renew our sense of wonder: the world is reconfigured and bright again. Meaning we are new. Meaning the meaning we're making can be too. Until we habituate and are old again, repeat, repeat.

And the best transport is—oddly, suddenly—the river. What was formerly gap is now

What you find in the bed of the river is another world:

Ecosystem for the vanished and sometimes the present, home for the homeless. The gutter is a place for the placeless teens in-between one world and another (the word teens connotes this betweenness even as we deify it in retrospect): of course they've found their way here and made their marks on walls:

And if these structures fail?

It's hard to ice over a dry river, though it's not hard to drown in one when it runs. Occasionally you might tube or kayak in one, but the fear here is flash flood and sudden rush, the quick filling-in of what you thought before was just a blank, the sudden reconfiguration of what you thought was your city, the roads suddenly water, the city no longer for driving, it's obvious.

solid, and what was solid, now absence.

beer cans; a couple hundred types of plants, only a few of which I can name; desert animals, confused or crafty; the supersonic clicks of migrating bats hanging under bridges and hunting by night; the occasional runner; the bat-watchers; the kids; graffiti; horses; those walking their pets. Shopping carts are my most obvious and often sighting—often from the nearby Lowe's or the nearby Fry's or the nearby Albertson's or the nearby Walmart or the nearby Costco or the nearby Supertarget, or the nearby whatever, abandoned here by teens, the homeless, or homeless teens, then found with a sigh some days hence. Some of it's pilfered by the river when it runs, but most of the human trash I find in the riverbed is abandoned here or carted in by squatters who carve their own spaces—campsites almost—out of the plants about midway up the riverbed. These spaces are not visible from the street or the walkways that run along the river. It's surprising when you run across one, and can only be accessed via the river: world within world within world.

Gateway to Hell —>
Get up off ya ass and wash it
Thank You Bitch: Heather Rose
Cream & Impulse, Lost Soul
Beware of Killer Frogs
Dan & Dalik, I Heart You, 5-28-91
Tom Died Here
Point of No Return
I live —>
OK N 2 DEEP
Fuck Life

Most interesting to me are the absences in the absences: the darkened tunnels that feed into the wash or the Rillito River. These storm sewers are somewhat rare in Tucson, especially the big ones, 72" wide and high, that I have been exploring, mapping, and annotating (where I found the graffiti above). They lead into the dark space

You never answered the question: is a river still a river if it's dry?

If we strip the river of river, if we de-wild the wild, if we fuck with—excuse me, engineer—the river enough it will certainly find a different route when it needs to. Sure, pave it over. Develop its banks. It'll tear all that up. It'll take over Swan and Campbell, main thoroughfares, making your Bruegger's Bagels, your Bed, Bath & Beyonds, your Chili's, sites of your future water landing. It will erase whatever is in the path of least resistance. The best you can do, the engineers say, is to try to control it, confine it.

Sometimes the river will do its river thing regardless. There's not much that can be done.

No *river* without *I*.

No *meander* without *ander*.

underneath the city and into the metaphorical darkness underneath our civilized, engineered lives. They too are colonized mostly by animals, teens, and the homeless, judging from the graffiti and the trash.

In 2008 the Santa Cruz River (that the Pantano and Rillito feed into) was ruled "navigable" by the Environmental Protection Agency, thus making it eligible for protection under the federal Clean Water Act. This followed the Army Corps of Engineers' second-guessing of their own ruling of "navigable" the previous May, then withdrawn, in effect punting to the EPA. "Navigable" is a tricky word, especially for the 90–95% of Arizona rivers that are dry for at least part of the year. Calling a river "navigable" means more roadblocks for developers wanting to build along the river. Calling a river "navigable" enshrines it, embodies it as a river deserving administrative, environmental protections. Calling a river "navigable" only if it runs year-round brands a western river with the mark of the midwest or the east. These notions don't really apply here.

The riverbed is certainly navigable in the dry months, just not by boat, by floating, by ogling the nonwet world from the center of the stream and trying not to get caught on debris.

(I know it sounds crazy, but perhaps we could relinquish some of that control?)

Sometimes the self will do its self thing regardless and spill into the floodplain.

Well, that's self-centered, and we're already well into the margin. No *in* either. No *either*. No *margin*, no *aquifer*, no *Pima*. No *Rillito*. And certainly no *memoir*, no *creative nonfiction*, no *running*. *I* is central to the river linguistically, but also politically. It's easy to offload, subcontract my concern, make it someone else's problem. And it is. It's not my river. It's not *just* my river. Though I suck at proselytizing, at conserving, it's obvious that Arizona is not Michigan, though it might be the distant future of Michigan once everything else is mined or sold. Given the water situation in the west, I bought rain barrels to harvest rainwater. Got solar. I'm lawnless, thank god, thank allergies, thank desert landscaping, thank the culture of lawnlessness built on what was once the lawless west, like Girls, gone Wild, gone Havasu, gone downloadable porno. Gone snowbird and gunmad and darkness. Gone bagless, teabagless. Gone badlands. Gone cholla, gone cactus, succulent, ocotillo. The fact of the desert is a sort of purification by sunlight and drought, and to understand that is to start to make a life here.

No *ander* without *other*. No *Duran Duran* without *ander ander* (say it enough & you'll get it). It's easy to get obsessed with linguistic tics like this. Do they mean or do they not? Do they make the world about us or do they not?

They mean if we make them mean. Meaning will fill any space we can carve.

But if left unchecked it becomes a problem.

Meaninglessness and lawnlessness and selflessness do not stop my neighbors from trying to involve me in their yearly competitive masculine yard undertakings, but at least I am no longer a slave to the neighborhood organization and the uncut lawn citations that I dutifully composted.

To be lawnless is to opt out of that one cultural norm. From the dusty Tucson margin, ridiculous, lawned Phoenix and the faux-Venetian Scottsdale are easy targets for critique or meandering rage—

And while I may be asymptotically approaching middle age, my running in the river will beat back time, slow the growth of the blood plaque in my arterial tributaries as I dream cold rushing something in the Pantano, the Rillito, the Santa Cruz as I map and annotate its storm drain tributaries. It's a new interface with the ground below me that I've found powerful and surprising and perhaps empowering, though it's hard to say it that way these days; either I'm newer or it's newer, or both. This is how I believe the I can feel, running along the Rillito on overcast days just edging the darkness of self and of blindness and of nearly infinite complication.

Does thinking change anything, I wonder?

I wander—meander—the river, this river, this weather, this sputtering gutter,

We have no option but to apprehend the world through the lens of self, the double heart of landscape we carry with us.

The problem of memoir. The problem of me. The problem of Facebook. The problem of victim culture. Of Not in My Backyard, of apathy and anger. The problem of no vision, of television, of unsupervised time with teens we lust for but don't trust.

Living in Tucson is being reminded of lawlessness, of lawnlessness, of the edge of the culture past which we might drop off into nothingness. Pointless national television commercials tempt me with discounted riding lawnmowers offering tighter turning radii each spring. The lush, trimmed lawn is still the culture in much of the country. Our yards have been our cultures for centuries, probably since the earliest yard got spruced up. That we can control our exteriors suggests the straight cleanliness and orderliness of our souls, not to mention our manscaped chests. It's obvious, isn't it, those Sunday afternoons spent mowing after church, trimming the unruly, making a façade of control—the starch shaking out of our church shirts with every vibration, the clutch in the riding lawnmower getting looser by the day as we ride toward the future in the fading light, spitting gravel into the window when we hit it. We don't have that here and I'm thankful. Years of summers spent pointlessly mowing I'm thankful. Years of uselessly beating back wildness I'm thankful. Years of taming the rivers I'm thankful. Years of asthma attacks and allergy shots and shirtless underpaid work for middle-aged women who I may or may not have lusted for I'm thankful.

but not today, I think. This is about the river and the I-shape that the gutter and margin make that separates the pages, that separates thought from action, self from subject, past from future.

Does remembering?

dividing, connecting, eliding.

"I in River" first appeared in the journal *Passages North*.

Past Year's Data/Future Archive 2010–11

CHRISTIANA CARO

MFA, Photography, School of Art, 2011

Interdisciplinary Seminar: Critical Issues in Art + Design, 2010

I walked into a cave on the side of a rock face. "How we make our rocks." This exhibit gave it away without telling anything.

Stuffed animal, hunted, displayed. Am I looking at something dead or not dead?

Outside of the rock face I photographed an awkward teenaged boy dressed all in black. He carried an oversized rainbow umbrella (it is sunny in the desert). So I asked him to twirl it.

The desert was the sea. I am standing at the bottom of the ocean.

The Sky Island region of Southern Arizona encompasses multiple mountain ranges in the Rillito River valley that appear to float on the horizon when seen from elevated vantage points. I went in search of these mountaintop islands while shooting photographs along the way, in addition to making images in museums on the ground. Sky islands sought: The Huachucas, Santa Ritas, Pinaleños, Rincons, Dragoons, and Chiricahuas. Museums: The Arizona-Sonora Desert Museum, The International Wildlife Museum, Tucson's Western film sets, and the Mini Time Machine Museum of Miniatures, among others.

A river—that constant on the land that winds through valleys, across borders. The now-dry Rillito riverbed traces a path here, serving as a routed platform for this perceptual landscape. One cannot go to a sky island, just as one can never reach the horizon. Similarly, we cannot experience nature inside a room, and a river does not need to contain water to be defined as such. Contradictions, new ways of seeing, of looking.

This desert river swallows the sky's double, rendering a scar rather than a mirror on the land. The transformation of nature into culture was a logical counterpoint to the Rillito River's inverted reflection of its sky islands above. I travelled from ground to clouds, and then to the representation of these natural sites inside museums.

The ocean–desert tells a perverse ecological story as the Rillito withers more with each season. Here in the Sonoran Desert, illusory islands must rest in the sky, above and beyond the horizon, because its water has vanished. The sky island is a story, an unreachable myth that leads us to look and then vanishes again. The River, even dry, also has stories, also renders a line to be followed, and then too disappears before filling, briefly, then emptying again.

OPPOSITE PAGE *Corner* | digital c-type print | 6″ x 6″ | 2011

FOLLOWING PAGES *Flion* (diptych) | digital c-type print | 10″ x 20″ | 2011

PUSH
PUSH

BIOGRAPHIES

Jonathan Black, a graphic designer specializing in environmentally responsible design, received his MFA in Visual Communications from The University of Arizona in 2012 and is now co-owner of the Cast Iron Design Company based in Boulder, Colorado.

Alison Hawthorne Deming, UA Professor of Creative Writing, is the author of four poetry books including *Rope* (Penguin, 2009) and four books of nonfiction, including the forthcoming *Zoologies: On Animals and the Human Spirit*. She was Poet-in-Residence for the Language of Conservation project at the Jacksonville Zoo and Gardens in Florida.

Javier D. Duran, Director of Confluencenter for Creative Inquiry and UA Professor of Spanish and Border Studies, is the author of *José Revueltas: Una poética de la Disidencia* and has co-edited six books including *En la Línea/On the Line*. He would like to thank Confluencenter team members William Costley and Yvonne Ervin who provided intellectual inspiration and editorial guidance for his Publisher's Note.

Allison Dushane, UA Assistant Professor of Literature in the English Department, is currently working on a scholarly edition of Erasmus Darwin's two-part epic poem, *The Botanic Garden*, and finishing a manuscript titled *Autonomous Life: Matter and Agency in the Romantic Era*.

Gregg Garfin is Deputy Director for Science Translation and Outreach in the Institute of the Environment and UA Assistant Professor in the School of Natural Resources and the Environment. His research involves climate impacts on society, drought, climate change, and the use of climate information in resource management decision making.

Katharine L. Jacobs, UA Professor and Director of the National Climate Assessment in the White House, researches climate adaptation, water management issues, and science and decision making. She was the Director of the Arizona Water Institute and has 23 years of experience managing water resources in the Tucson, Arizona area.

Kelly Leslie, UA Associate Professor and Division Chair of Visual Communications in the School of Art, is a graphic designer, artist, and educator. She exhibits her drawings and prints widely and teaches typography, web design, and professional practices in design.

Ellen McMahon is a Fulbright Scholar and UA Professor of Art and Visual Communications. Her interest in combining the perspectives and methods of artists, designers, and scientists has led her into several collaborative projects such as this.

Ander Monson is the author of a number of paraphernalia including a website <otherelectricities.com>, a decoder wheel, several chapbooks, as well as five books, most recently *Vanishing Point: Not a Memoir* (Graywolf Press, 2010) and *The Available World* (Sarabande Books, 2010). He teaches in the UA Creative Writing MFA program and edits the journal *DIAGRAM* <thediagram.com> and New Michigan Press.

Logan Phillips is a bilingual poet, performer, teacher, and DJ who tours his art throughout the United States and internationally. Phillips works to bring poems to the stage with the transdisciplinary performance group Verbo•bala and is the author of the book *The Sonoran Strange*. Born in the borderlands, he lives in Tucson.

The Rillito River Project inaugurated in January 2007 by Creative Director Ellen Skotheim, was envisioned as a series of artworks to be installed or performed in the Rillito riverbed. Skotheim and Technical Director John Newman serve on the board of directors along with Ruth Bunny Davis, Paul Durham, Susan Lobo, Stephen Goldstein, and Susan Tarrence, who collectively define the mission of the Rillito River Project. Jennifer Powers-Murphy is the publicist.

Paul Robbins is Director of the Nelson Institute for Environmental Studies at the University of Wisconsin and is author of several books, including *Lawn People*. He would like to thank Dr. Sarah Moore, whose inspiration on the power of objects on the psyche was indispensable to his conclusions concerning team research.

Beth Weinstein is an architect, Associate Professor in the UA School of Architecture, and performance scholar and designer. She curated the *Collaborative Legacy of Merce Cunningham* exhibition and works on projects at the seam between architecture and choreography, and that prioritize connections between things over the things themselves.